# Anatomy of the Ship

# Anatomy of the Ship

Shiven Arora

RANDOM

RANDOM PUBLICATIONS
NEW DELHI (INDIA)

**Anatomy of the Ship**

ISBN 978-93-5111-401-7


Published in 2014 in India by

**RANDOM PUBLICATIONS**

4376-A/4B, Gali Murari Lal, Ansari Road
New Delhi-110 002
Phone : +91-11-43580356, +91-11-23289044
e-mail: randomexports@gmail.com, sales@randompublications.com, info@randompublications.com

*Type Setting by* : Keystoneprintads, Delhi-110051
*Printed at* : Mehra Printers, Delhi-110 092

# Preface

Since the end of the age of sail a ship has been any large buoyant watercraft. Ships are generally distinguished from boats based on size, shape and cargo or passenger capacity. Ships are used on lakes, seas, and rivers for a variety of activities, such as the transport of people or goods, fishing, entertainment, public safety, and warfare. Historically, a "ship" was a vessel with sails rigged in a specific manner. Ships and boats have developed alongside humanity. In armed conflict and in daily life they have become an integral part of modern commercial and military systems.

Fishing boats are used by millions of fishermen throughout the world. Military forces operate vessels for combat and to transport and support forces ashore. Commercial vessels, nearly 35,000 in number, carried 7.4 billion tons of cargo in 2007. Total number of ships as of 2011 is about 104,304. Ships were a key in history's great explorations and scientific and technological development. Navigators such as Zheng He spread such inventions as the compass and gunpowder.

Ships have been used for such purposes as colonization and the slave trade, and have served scientific, cultural, and humanitarian needs. After the 16th century, new crops that had come from and to the Americas via the European seafarers significantly contributed to the world's population growth. Maritime transport has shaped the world's economy into today's energy-intensive pattern.

Some components exist in vessels of any size and purpose. Every vessel has a hull of sorts. Every vessel has some sort of propulsion, whether it's a pole, an ox, or a nuclear reactor. Most vessels have some sort of steering system. Other characteristics are common, but not as universal, such as compartments, holds, a superstructure, and equipment such as anchors and winches. For a ship to float, its weight must be less than that of the water displaced by the ship's hull. There are many types of hulls, from logs lashed together to form a raft to the advanced hulls of America's Cup sailboats. A vessel may have a single hull, two in the case of catamarans, or three in the case of trimarans. Vessels with more than three hulls are rare, but some experiments have been conducted with designs such as pentamarans. Multiple hulls are generally parallel to each other and connected by rigid arms. Hulls have several elements. The bow is the foremost part of the hull. Many ships feature a bulbous bow. The keel is at the very bottom of the hull, extending the entire length of the ship. The rear part of the hull is known as the stern, and many hulls have a flat back known as a transom. Common hull appendages include propellers for propulsion, rudders for steering, and stabilizers to quell a ship's rolling motion.

Other hull features can be related to the vessel's work, such as fishing gear and sonar domes. Hulls are subject to various hydrostatic and hydrodynamic constraints. The key hydrostatic constraint is that it must be able to support the entire weight of the boat, and maintain stability even with often unevenly distributed weight. Hydrodynamic constraints include the ability to withstand shock waves, weather collisions and groundings. Older ships and pleasure craft often have or had wooden hulls. Steel is used for most commercial vessels. Aluminium is frequently used for fast vessels, and composite materials are often found in sailboats and pleasure craft. Some ships have been made with concrete hulls.

I thank all members of my team who have helped in the preparation of the book. My special thanks go to "Random Publications" who have published the book.

– ***Shiven Arora***

# Contents

# 1

# Ship: An Introduction

Since the end of the age of sail a ship has been any large buoyant watercraft. Ships are generally distinguished from boats based on size and cargo or passenger capacity. Ships are used on lakes, seas, and rivers for a variety of activities, such as the transport of people or goods, fishing, entertainment, public safety, and warfare. Historically, a "ship" was a vessel with sails rigged in a specific manner.

Ships and boats have developed alongside humanity. In armed conflict and in daily life they have become an integral part of modern commercial and military systems. Fishing boats are used by millions of fishermen throughout the world. Military forces operate vessels for combat and to transport and support forces ashore. Commercial vessels, nearly 35,000 in number, carried 7.4 billion tons of cargo in 2007. Total number of ships(with IMO number) as of 2011 is about 104,304.

Ships were a key in history's great explorations and scientific and technological development. Navigators such as Zheng He spread such inventions as the compass and gunpowder. Ships have been used for such purposes as colonization and the slave trade, and have served scientific, cultural, and humanitarian needs. After the 16th century, new crops that had come from and to the Americas via the European seafarers significantly contributed to the world's population growth. Maritime transport has shaped the world's economy into today's energy-intensive pattern.

## NOMENCLATURE

Ships can usually be distinguished from boats based on size and the ship's ability to operate independently for extended periods. A commonly used rule of thumb is that if one vessel can carry another, the larger of the two is a ship. Dinghies are carried on sailing yachts as small as 35 feet (10.67 m), clearly not ships; this rule of thumb is not foolproof.

In the age of sail, a "ship" was a sailing vessel with at least three square-rigged masts and a full bowsprit; other types of vessel were also defined by their sailplan, e.g. barque, brigantine, etc. A number of large vessels are usually referred to as boats. Submarines are a prime example. Other types of large vessel which are traditionally called boats are the Great Lakes freighter, the riverboat, and the ferryboat. Though large enough to carry their own boats and heavy cargoes, these vessels are designed for operation on inland or protected coastal waters.

In most maritime traditions ships have individual names, and modern ships may belong to a ship class often named after its first ship. In English, a ship is traditionally referred to as "she", even if named after a man, but this is not universal usage; some journalistic style guides advise using "it", others advise "she" and "her".

## HISTORY

### PREHISTORY AND ANTIQUITY

The first known vessels date back to the Neolithic Period, about 10,000 years ago, but could not be described as ships. The first navigators began to use animal skins or woven fabrics as sails. Affixed to the top of a pole set upright in a boat, these sails gave early ships range. This allowed men to explore widely, allowing for the settlement of Oceania for example (about 3,000 years ago).

By around 3000 BC, Ancient Egyptians knew how to assemble wooden planks into a hull. They used woven straps to lash the planks together, and reeds or grass stuffed between the planks helped to seal the seams. The Greek historian and geographer Agatharchides had documented ship-faring among the early Egyptians: *"During the prosperous period of the Old Kingdom, between the 30th and 25th centuries B. C., the river-routes were kept in order, and Egyptian ships sailed the Red Sea as far as the myrrh-country."* Sneferu's ancient cedar wood ship Praise of the Two Lands is the first reference recorded (2613 BCE) to a ship being referred to by name.

The ancient Egyptians were perfectly at ease building sailboats. A remarkable example of their shipbuilding skills was the Khufu ship, a vessel 143 feet (44 m) in length entombed at the foot of the Great Pyramid of Giza around 2500 BC and found intact in 1954.

It is known that ancient Nubia/Axum traded with India, and there is evidence that ships from Northeast Africa may have sailed back and forth between India/Sri Lanka and Nubia trading goods and even to Persia, Himyar and Rome. Aksum was known by the Greeks for having seaports for ships from Greece and Yemen.

Elsewhere in Northeast Africa, the Periplus of the Red Sea reports that Somalis, through their northern ports such as Zeila and Berbera, were trading frankincense and other items with the inhabitants of the Arabian Peninsula well before the arrival of Islam as well as with then Roman-controlled Egypt.

A panel found at Mohenjodaro depicted a sailing craft. Vessels were of many types; their construction is vividly described in the Yukti Kalpa Taru, an ancient Indian text on shipbuilding. This treatise gives a technical exposition on the techniques of shipbuilding. It sets forth minute details about the various types of ships, their sizes, and the materials from which they were built. The Yukti Kalpa Taru sums up in a condensed form all the available information. The Yukti Kalpa Taru gives sufficient information and dates to prove that, in ancient times, Indian shipbuilders had a good knowledge of the materials which were used in building ships.

In addition to describing the qualities of the different types of wood and their suitability for shipbuilding, the Yukti Kalpa Taru gives an elaborate classification of ships based on their size.

By about 2000 BC, the Minoan civilization in Crete had evolved into a naval power exercising effective control of the sea in the eastern Mediterranean.

The Phoenicians, the first to sail completely around Africa, and Greeks gradually mastered navigation at sea aboard triremes, exploring and colonizing the Mediterranean via ship. Around 340 BC, the Greek navigator Pytheas of Massalia ventured from Greece to Western Europe and Great Britain. In the course of the 2nd century BC, Rome went on to destroy Carthage and subdue the Hellenistic kingdoms of the eastern Mediterranean, achieving complete mastery of the inland sea, that they called *Mare Nostrum*. The monsoon wind system of the Indian Ocean was first sailed by Greek navigator Eudoxus of Cyzicus in 118 BC. With 300 Greek ships a year sailing between Roman Empire and India, the annual trade may have reached 300,000 tons.

In China, by the time of the Zhou Dynasty ship technologies such as stern mounted rudders were developed, and by the Han Dynasty, a well kept naval fleet was an integral part of the military. Ship technology advanced to the point where by the medieval period, water tight compartments were developed.

The Swahili people had various extensive trading ports dotting the coast of medieval East Africa and Great Zimbabwe had extensive trading contacts with Central Africa, and likely also imported goods brought to Africa through the Southeast African shore trade of Kilwa in modern-day Tanzania.

It is known by historians that at its height the Mali Empire built a large naval fleet under Emperor Mansa Musa in the late 13th and early 14th century. Arabic sources describe what some consider to be visits to the New World by a Mali fleet in 1311.

Before the introduction of the compass, celestial navigation was the main method for navigation at sea. In China, early versions of the magnetic compass were being developed and used in navigation between 1040 and 1117. The true mariner's compass, using a pivoting needle in a dry box, was developed in Europe no later than 1300.

## RENAISSANCE

Until the Renaissance, navigational technology remained comparatively primitive. This absence of technology did not prevent some civilizations from becoming sea powers. Examples include the maritime republics of Genoa and Venice, Hanseatic League, and the Byzantine navy. The Vikings used their knarrs to explore North America, trade in the Baltic Sea and plunder many of the coastal regions of Western Europe.

Towards the end of the 14th century, ships like the carrack began to develop towers on the bow and stern. These towers decreased the vessel's stability, and in the 15th century, the caravel, designed by the Portuguese, based on the Arabic *qarib* which could sail closer to the wind, became more widely used. The towers were gradually replaced by the forecastle and sterncastle, as in the carrack *Santa María* of Christopher Columbus. This increased freeboard allowed another innovation: the freeing port, and the artillery associated with it.

In the 16th century, the use of freeboard and freeing ports become widespread on galleons. The English modified their vessels to maximize their firepower and demonstrated the effectiveness of their doctrine, in 1588, by defeating the Spanish Armada.

At this time, ships were developing in Asia in much the same way as Europe. Japan used defensive naval techniques in the Mongol invasions of Japan in 1281. It is likely that the Mongols of the time took advantage of both European and Asian shipbuilding techniques. During the 15th century, China's Ming Dynasty assembled one of the largest and most powerful naval fleet in the world for the diplomatic and power projection voyages of Zheng He. Elsewhere in Japan in the 15th century, one of the world's first iron-clads, "Tekkôsen", literally meaning "iron ships", was also developed. In Japan, during the Sengoku era from the fifteenth to 17th century, the great struggle for feudal supremacy was fought, in part, by coastal fleets of several hundred boats, including the atakebune.

During the Age of the Ajuuraan, the Somali sultanates and republics of Merca, Mogadishu, Barawa, Hobyo and their respective ports flourished, enjoying a lucrative foreign commerce with ships sailing to and coming from Arabia, India, Venetia, Persia, Egypt, Portugal and as far away as China. In the 16th century, Duarte Barbosa noted that many ships from the Kingdom of Cambaya in what is modern-day India sailed to Mogadishu with cloth and spices, for which they in return received gold, wax and ivory. Barbosa also highlighted the abundance of meat, wheat, barley, horses, and fruit on the coastal markets, which generated enormous wealth for the merchants.

Middle Age Swahili Kingdoms are known to have had trade port islands and trade routes with the Islamic world and Asia and were described by Greek historians as "metropolises". Famous African trade ports such as Mombasa, Zanzibar, and Kilwa were known to Chinese sailors such as Zheng He and medieval Islamic historians such as the Berber Islamic voyager Abu Abdullah ibn Battua. In the 14th century CE King Abubakari I, the brother of King Mansa Musa of the Mali Empire is thought to have had a great armada of ships sitting on the coast of West Africa. This is corroborated by ibn Battuta himself who recalls several hundred Malian ships off the coast. This has led to great speculation, with historical evidence, that it is possible that Malian sailors may have reached the coast of Pre-Columbian America under the rule of Abubakari II, nearly two hundred years before Christopher Columbus and that black traders may have been in the Americas before Columbus.

Fifty years before Christopher Columbus, Chinese navigator Zheng He traveled the world at the head of what was for the time a huge armada. The largest of his ships had nine masts, were 130 metres (430 ft) long and had a beam of 55 metres (180 ft). His fleet carried 30,000 men aboard 70 vessels, with the goal of bringing glory to the Chinese emperor.

The carrack and then the caravel were developed in Iberia. After Columbus, European exploration rapidly accelerated, and many new trade routes were established. In 1498, by reaching India, Vasco da Gama proved that the access to the Indian Ocean from the Atlantic was possible. These explorations in the Atlantic and Indian Oceans were soon followed by France, England and the Netherlands, who explored the Portuguese and Spanish trade routes into the Pacific Ocean, reaching Australia in 1606 and New Zealand in 1642. A major sea power, the Dutch in 1650 owned 16,000 merchant ships. In the 17th century Dutch explorers such as Abel Tasman explored the coasts of Australia, while in the 18th century it was British explorer James Cook who mapped much of Polynesia.

## SPECIALIZATION AND MODERNIZATION

Parallel to the development of warships, ships in service of marine fishery and trade also developed in the period between antiquity and the Renaissance. Still primarily a coastal endeavor, fishing is largely practiced by individuals with little other money using small boats.

Maritime trade was driven by the development of shipping companies with significant financial resources. Canal barges, towed by draft animals on an adjacent towpath, contended with the railway up to and past the early days of the industrial revolution. Flat-bottomed and flexible scow boats also became widely used for transporting small cargoes. Mercantile trade went hand-in-hand with exploration, self-financed by the commercial benefits of exploration.

During the first half of the 18th century, the French Navy began to develop a new type of vessel known as a ship of the line, featuring seventy-four guns. This type of ship became the backbone of all European fighting fleets. These ships were 56 metres (184 ft) long and their construction required 2,800 oak trees and 40 kilometres (25 mi) of rope; they carried a crew of about 800 sailors and soldiers.

During the 19th century the Royal Navy enforced a ban on the slave trade, acted to suppress piracy, and continued to map the world. A clipper was a very fast sailing ship of the 19th century. The clipper route fell into commercial disuse with the introduction of steam ships, and the opening of the Suez and Panama Canals.

Ship designs stayed fairly unchanged until the late 19th century. The industrial revolution, new mechanical methods of propulsion, and the ability to construct ships from metal triggered an explosion in ship design. Factors including the quest for more efficient ships, the end of long running and wasteful maritime conflicts, and the increased financial capacity of industrial powers created an avalanche of more specialized boats and ships. Ships built for entirely new functions, such as firefighting, rescue, and research, also began to appear.

In light of this, classification of vessels by type or function can be difficult. Even using very broad functional classifications such as fishery, trade, military, and exploration fails to classify most of the old ships. This difficulty is increased by the fact that the terms such as sloop and frigate are used by old and new ships alike, and often the modern vessels sometimes have little in common with their predecessors.

### TODAY

In 2007, the world's fleet included 34,882 commercial vessels with gross tonnage of more than 1,000 tons, totaling 1.04 billion tons. These ships carried 7.4 billion tons of cargo in 2006, a sum that grew by 8% over the previous year. In terms of tonnage, 39% of these ships are tankers, 26% are bulk carriers, 17% container ships and 15% were other types.

In 2002, there were 1,240 warships operating in the world, not counting small vessels such as patrol boats. The United States accounted for 3 million tons worth of these vessels, Russia 1.35 million tons, the United Kingdom 504,660 tons and China 402,830 tons. The 20th century saw many naval engagements during the two world wars, the Cold War, and the rise to power of naval forces of the two blocs. The world's major powers have recently used their naval power in cases such as the United Kingdom in the Falkland Islands and the United States in Iraq.

The size of the world's fishing fleet is more difficult to estimate. The largest of these are counted as commercial vessels, but the smallest are legion. Fishing vessels can be found in most seaside villages in the world. As of 2004, the United Nations Food and Agriculture Organization estimated 4 million fishing vessels were operating worldwide. The same study estimated that the world's 29 million fishermen caught 85,800,000 tonnes (84,400,000 long tons; 94,600,000 short tons) of fish and shellfish that year.

## TYPES OF SHIPS

Ships are difficult to classify, mainly because there are so many criteria to base classification on. One classification is based on propulsion; with ships categorised as a sailing ship, a steamship, or a motorship. Sailing ships are propelled solely by means of sails. Steamships are propelled by steam engines. Motorships use internal combustion engines; they include ships propelled by a combination of sail and internal combustion.

Ships can also be classified by other criteria such as:

- The number of hulls: monohull, catamaran, trimaran.
- The shape, size, and function, giving categories such as dinghy, keelboat, and icebreaker.
- The hull material: steel, aluminum, wood, fiberglass, and plastic.
- The type of propulsion system used, giving human-propelled (e.g., historical triremes), mechanical, and sails.
- The epoch in which the vessel was used, triremes of Ancient Greece, ships of the line of battle in the 18th century.
- The geographic origin of the vessel; many vessels are associated with a particular region, such as the pinnace of Northern Europe, the gondolas of Venice, and the junks of China.
- The manufacturer, series, or class.

Another way to categorize ships and boats is based on their use, as described by Paulet and Presles. This system includes military ships, commercial vessels, fishing boats, pleasure craft and competitive boats. In this section, ships are classified using the first four of those categories, and adding a section for lake and river boats, and one for vessels which fall outside these categories.

## COMMERCIAL VESSELS

Commercial vessels or merchant ships can be divided into three broad categories: cargo ships, passenger ships, and special-purpose ships. Cargo ships transport dry and liquid cargo. Dry cargo can be transported in bulk by bulk carriers, packed directly onto a general cargo ship in break-bulk, packed in intermodal containers as aboard a container ship, or driven aboard as in roll-on roll-off ships.

Liquid cargo is generally carried in bulk aboard tankers, such as oil tankers which may include both crude and finished products of oil, chemical tankers which may also carry vegetable oils other than chemicals and LPG/ LNG tankers, although smaller shipments may be carried on container ships in tank containers.

Passenger ships range in size from small river ferries to very large cruise ships. This type of vessel includes ferries, which move passengers and vehicles on short trips; ocean liners, which carry passengers from one place to another; and cruise ships, which carry passengers on voyages undertaken for pleasure, visiting several places and with leisure activities on board, often returning them to the port of embarkation. Special-purpose vessels are not used for transport but for other tasks. Examples include tugboats, pilot boats, rescue boats, cable ships, research vessels, survey vessels, and icebreakers. Most commercial vessels have full hull-forms to maximize cargo capacity. Hulls are usually made of steel, although aluminum can be used on faster craft, and fiberglass on the smallest service vessels. Commercial vessels generally have a crew headed by a captain, with deck officers and marine engineers on larger vessels. Special-purpose vessels often have specialized crew if necessary, for example scientists aboard research vessels.

Commercial vessels are typically powered by a single propeller driven by a diesel or, less usually, gas turbine engine. The fastest vessels may use pump-jet engines.

## NAVAL VESSELS

Naval vessels are those used by a navy for military purposes. There have been many types of naval vessel. Modern naval vessels can be broken down into three categories: surface warships, submarines, and support and auxiliary vessels.

Modern warships are generally divided into seven main categories: aircraft carriers, cruisers, destroyers, frigates, corvettes, submarines and amphibious assault ships. The distinction between cruisers, destroyers, frigates, and corvettes is not rigorous; the same vessel may be described differently in different navies. Battleships were used during the Second World War and occasionally since then (the last battleships were removed from the U.S. Naval Vessel Register in March 2006), but were made obsolete by the use of carrier-borne aircraft and guided missiles.

Most military submarines are either attack submarines or ballistic missile submarines. Until the end of World War II the primary role of the diesel/electric submarine was anti-ship warfare, inserting and removing covert agents and military forces, and intelligence-gathering. With the development of the homing torpedo, better sonar systems, and nuclear propulsion, submarines also became able to effectively hunt each other. The development of submarine-launched nuclear and cruise missiles gave submarines a substantial and long-ranged ability to attack both land and sea targets with a variety of weapons ranging from cluster bombs to nuclear weapons.

Most navies also include many types of support and auxiliary vessel, such as minesweepers, patrol boats, offshore patrol vessels, replenishment ships, and hospital ships which are designated medical treatment facilities.

Fast combat vessels such as cruisers and destroyers usually have fine hulls to maximize speed and maneuverability. They also usually have advanced electronics and communication systems, as well as weapons.

## FISHING VESSELS

Fishing vessels are a subset of commercial vessels, but generally small in size and often subject to different regulations and classification. They can be categorized by several criteria: architecture, the type of fish they catch, the fishing method used, geographical origin, and technical features such as rigging.

As of 2004, the world's fishing fleet consisted of some 4 million vessels. Of these, 1.3 million were decked vessels with enclosed areas and the rest were open vessels. Most decked vessels were mechanized, but two-thirds of the open vessels were traditional craft propelled by sails and oars. More than 60% of all existing large fishing vessels were built in Japan, Peru, the Russian Federation, Spain or the United States of America.

Fishing boats are generally small, often little more than 30 meters (98 ft) but up to 100 metres (330 ft) for a large tuna or whaling ship. Aboard a fish processing vessel, the catch can be made ready for market and sold more quickly once the ship makes port. Special purpose vessels have special gear. For example, trawlers have winches and arms, stern-trawlers have a rear ramp, and tuna seiners have skiffs.

In 2004, 85,800,000 tonnes (84,400,000 long tons; 94,600,000 short tons) of fish were caught in the marine capture fishery. Anchoveta represented the largest single catch at 10,700,000 tonnes (10,500,000 long tons; 11,800,000 short tons). That year, the top ten marine capture species also included Alaska pollock, Blue whiting, Skipjack tuna, Atlantic herring, Chub mackerel, Japanese anchovy, Chilean jack mackerel, Largehead hairtail, and Yellowfin tuna. Other species including salmon, shrimp, lobster, clams, squid and crab, are also commercially fished.

Modern commercial fishermen use many methods. One is fishing by nets, such as purse seine, beach seine, lift nets, gillnets, or entangling nets. Another is trawling, including bottom trawl. Hooks and lines are used in methods like long-line fishing and hand-line fishing. Another method is the use of fishing trap.

## WEATHER VESSELS

A weather ship was a ship stationed in the ocean as a platform for surface and upper air meteorological observations for use in marine weather forecasting. Surface weather observations were taken hourly, and four radiosonde releases occurred daily. It was also meant to aid in search and rescue operations and to support transatlantic flights. Proposed as early as 1927 by the aviation community, the establishment of weather ships proved to be so useful during World War II that the International Civil Aviation Organization (ICAO) established a global network of weather ships in 1948, with 13 to be supplied by the United States. This number was eventually negotiated down to nine.

Their crews were normally out to sea for three weeks at a time, returning to port for 10 day stretches. Weather ship observations proved to be helpful in wind and wave studies, as they did not avoid weather systems like other ships tended to for safety reasons. They were also helpful in monitoring storms at sea, such as tropical cyclones. The removal of a weather ship became a negative factor in forecasts leading up to the Great Storm of 1987. Beginning in the 1970s, their role became largely superseded by weather buoys due to the ships' significant cost.

The agreement of the use of weather ships by the international community ended in 1990. The last weather ship was *Polarfront*, known as weather station M ("Mike"), which was put out of operation on 1 January 2010. Weather observations from ships continue from a fleet of voluntary merchant vessels in routine commercial operation.

## INLAND AND COASTAL BOATS

Many types of boats and ships are designed for inland and coastal waterways. These are the vessels that trade upon the lakes, rivers and canals. Barges are a prime example of inland vessels. Flat-bottomed boats built to transport heavy goods, most barges are not self-propelled and need to be moved by tugboats towing or towboats pushing them.

Barges towed along canals by draft animals on an adjacent towpath contended with the railway in the early industrial revolution but were out competed in the carriage of high value items because of the higher speed, falling costs, and route flexibility of rail transport.

Riverboats and inland ferries are specially designed to carry passengers, cargo, or both in the challenging river environment. Rivers present special hazards to vessels. They usually have varying water flows that alternately lead to high speed water flows or protruding rock hazards. Changing siltation patterns may cause the sudden appearance of shoal waters, and often floating or sunken logs and trees (called snags) can endanger the hulls and propulsion of riverboats.

Riverboats are generally of shallow draft, being broad of beam and rather square in plan, with a low freeboard and high topsides. Riverboats can survive with this type of configuration as they do not have to withstand the high winds or large waves that are seen on large lakes, seas, or oceans.

Lake freighters, also called lakers, are cargo vessels that ply the Great Lakes. The most well-known is the SS *Edmund Fitzgerald*, the latest major vessel to be wrecked on the Lakes. These vessels are traditionally called boats, not ships. Visiting ocean-going vessels are called "salties." Because of their additional beam, very large salties are never seen inland of the Saint Lawrence Seaway. Because the smallest of the Soo Locks is larger than any Seaway lock, salties that can pass through the Seaway may travel anywhere in the Great Lakes. Because of their deeper draft, salties may accept partial loads on the Great Lakes, "topping off" when they have exited the Seaway.

Similarly, the largest lakers are confined to the Upper Lakes (Superior, Michigan, Huron, Erie) because they are too large to use the Seaway locks, beginning at the Welland Canal that bypasses the Niagara River. Since the freshwater lakes are less corrosive to ships than the salt water of the oceans, lakers tend to last much longer than ocean freighters. Lakers older than 50 years are not unusual, and as of 2005, all were over 20 years of age.

The *St. Mary's Challenger*, built in 1906 as the *William P Snyder*, is the oldest laker still working on the Lakes. Similarly, the *E.M. Ford*, built in 1898 as the *Presque Isle*, was sailing the lakes 98 years later in 1996. As of 2007 the *Ford* was still afloat as a stationary transfer vessel at a riverside cement silo in Saginaw, Michigan.

## ARCHITECTURE

Some components exist in vessels of any size and purpose. Every vessel has a hull of sorts. Every vessel has some sort of propulsion, whether it's a pole, an ox, or a nuclear reactor. Most vessels have some sort of steering system. Other characteristics are common, but not as universal, such as compartments, holds, a superstructure, and equipment such as anchors and winches.

### HULL

For a ship to float, its weight must be less than that of the water displaced by the ship's hull. There are many types of hulls, from logs lashed together to form a raft to the advanced hulls of America's Cup sailboats. A vessel may have a single hull (called a monohull design), two in the case of catamarans, or three in the case of trimarans.

Vessels with more than three hulls are rare, but some experiments have been conducted with designs such as pentamarans. Multiple hulls are generally parallel to each other and connected by rigid arms.

Hulls have several elements. The bow is the foremost part of the hull. Many ships feature a bulbous bow. The keel is at the very bottom of the hull, extending the entire length of the ship. The rear part of the hull is known as the stern, and many hulls have a flat back known as a transom. Common hull appendages include propellers for propulsion, rudders for steering, and stabilizers to quell a ship's rolling motion. Other hull features can be related to the vessel's work, such as fishing gear and sonar domes.

Hulls are subject to various hydrostatic and hydrodynamic constraints. The key hydrostatic constraint is that it must be able to support the entire weight of the boat, and maintain stability even with often unevenly distributed weight. Hydrodynamic constraints include the ability to withstand shock waves, weather collisions and groundings.

Older ships and pleasure craft often have or had wooden hulls. Steel is used for most commercial vessels. Aluminium is frequently used for fast vessels, and composite materials are often found in sailboats and pleasure craft. Some ships have been made with concrete hulls.

## PROPULSION SYSTEMS

Propulsion systems for ships fall into three categories: human propulsion, sailing, and mechanical propulsion. Human propulsion includes rowing, which was used even on large galleys. Propulsion by sail generally consists of a sail hoisted on an erect mast, supported by stays and spars and controlled by ropes. Sail systems were the dominant form of propulsion until the 19th century. They are now generally used for recreation and competition, although experimental sail systems, such as the turbosails, rotorsails, and wingsails have been used on larger modern vessels for fuel savings.

Mechanical propulsion systems generally consist of a motor or engine turning a propeller, or less frequently, an impeller or wave propulsion fins. Steam engines were first used for this purpose, but have mostly been replaced by two-stroke or four-stroke diesel engines, outboard motors, and gas turbine engines on faster ships. Nuclear reactors producing steam are used to propel warships and icebreakers, and there have been attempts to utilize them to power commercial vessels.

In addition to traditional fixed and controllable pitch propellers there are many specialized variations, such as contra-rotating and nozzle-style propellers. Most vessels have a single propeller, but some large vessels may have up to four propellers supplemented with transverse thrusters for maneuvring at ports.

The propeller is connected to the main engine via a propeller shaft and, in case of medium- and high-speed engines, a reduction gearbox. Some modern vessels have a diesel-electric powertrain in which the propeller is turned by an electric motor powered by the ship's generators.

## STEERING SYSTEMS

For ships with independent propulsion systems for each side, such as manual oars or some paddles, steering systems may not be necessary. In most designs, such as boats propelled by engines or sails, a steering system becomes necessary. The most common is a rudder, a submerged plane located at the rear of the hull.

Rudders are rotated to generate a lateral force which turns the boat. Rudders can be rotated by a tiller, manual wheels, or electro-hydraulic systems. Autopilot systems combine mechanical rudders with navigation systems. Ducted propellers are sometimes used for steering. Some propulsion systems are inherently steering systems. Examples include the outboard motor, the bow thruster, and the Z-drive. Some sails, such as jibs and the mizzen sail on a ketch rig, are used more for steering than propulsion.

## HOLDS, COMPARTMENTS, AND THE SUPERSTRUCTURE

Larger boats and ships generally have multiple decks and compartments. Separate berthings and heads are found on sailboats over about 25 feet (7.6

m). Fishing boats and cargo ships typically have one or more cargo holds. Most larger vessels have an engine room, a galley, and various compartments for work. Tanks are used to store fuel, engine oil, and fresh water. Ballast tanks are equipped to change a ship's trim and modify its stability.

Superstructures are found above the main deck. On sailboats, these are usually very low. On modern cargo ships, they are almost always located near the ship's stern. On passenger ships and warships, the superstructure generally extends far forward.

### EQUIPMENT

Shipboard equipment varies from ship to ship depending on such factors as the ship's era, design, area of operation, and purpose. Some types of equipment that are widely found include:

- Masts can be the home of antennas, navigation lights, radar transponders, fog signals, and similar devices often required by law.
- Ground tackle includes equipment such as mooring winches, windlasses, and anchors. Anchors are used to moor ships in shallow water. They are connected to the ship by a rope or chain. On larger vessels, the chain runs through a hawsepipe.
- Cargo equipment such as cranes and cargo booms are used to load and unload cargo and ship's stores.
- Safety equipment such as lifeboats, liferafts, and survival suits are carried aboard many vessels for emergency use.

## DESIGN CONSIDERATIONS

### HYDROSTATICS

Boats and ships are kept on (or slightly above) the water in three ways:

- For most vessels, known as displacement vessels, the vessel's weight is offset by that of the water displaced by the hull.
- For planing ships and boats, such as the hydrofoil, the lift developed by the movement of the foil through the water increases with the vessel's speed, until the vessel is foilborne.
- For non-displacement craft such as hovercraft and air-cushion vehicles, the vessel is suspended over the water by a cushion of high-pressure air it projects downwards against the surface of the water.

A vessel is in equilibrium when the upwards and downwards forces are of equal magnitude. As a vessel is lowered into the water its weight remains constant but the corresponding weight of water displaced by its hull increases. When the two forces are equal, the boat floats. If weight is evenly distributed throughout the vessel, it floats without trim or heel. A vessel's stability is considered in both this hydrostatic sense as well as a hydrodynamic sense, when subjected to movement, rolling and pitching, and the action of waves and wind. Stability problems can lead to excessive pitching and rolling, and eventually capsizing and sinking.

## HYDRODYNAMICS

The advance of a vessel through water is resisted by the water. This resistance can be broken down into several components, the main ones being the friction of the water on the hull and wave making resistance. To reduce resistance and therefore increase the speed for a given power, it is necessary to reduce the wetted surface and use submerged hull shapes that produce low amplitude waves.

To do so, high-speed vessels are often more slender, with fewer or smaller appendages. The friction of the water is also reduced by regular maintenance of the hull to remove the sea creatures and algae that accumulate there. Antifouling paint is commonly used to assist in this. Advanced designs such as the bulbous bow assist in decreasing wave resistance.

A simple way of considering wave-making resistance is to look at the hull in relation to its wake. At speeds lower than the wave propagation speed, the wave rapidly dissipates to the sides.

As the hull approaches the wave propagation speed, however, the wake at the bow begins to build up faster than it can dissipate, and so it grows in amplitude. Since the water is not able to "get out of the way of the hull fast enough", the hull, in essence, has to climb over or push through the bow wave. This results in an exponential increase in resistance with increasing speed.

This hull speed is found by the formula:

$$knots \approx 1.34 \times \sqrt{Lft}$$

or, in metric units:

$$knots \approx 2.5 \times \sqrt{Lm}$$

where *L* is the length of the waterline in feet or meters.

When the vessel exceeds a speed/length ratio of 0.94, it starts to outrun most of its bow wave, and the hull actually settles slightly in the water as it is now only supported by two wave peaks. As the vessel exceeds a speed/length ratio of 1.34, the hull speed, the wavelength is now longer than the hull, and the stern is no longer supported by the wake, causing the stern to squat, and the bow rise. The hull is now starting to climb its own bow wave, and resistance begins to increase at a very high rate. While it is possible to drive a displacement hull faster than a speed/length ratio of 1.34, it is prohibitively expensive to do so. Most large vessels operate at speed/length ratios well below that level, at speed/length ratios of under 1.0.

For large projects with adequate funding, hydrodynamic resistance can be tested experimentally in a hull testing pool or using tools of computational fluid dynamics. Vessels are also subject to ocean surface waves and sea swell as well as effects of wind and weather. These movements can be stressful for passengers and equipment, and must be controlled if possible. The rolling movement can be controlled, to an extent, by ballasting or by devices such as fin stabilizers. Pitching movement is more difficult to limit and can be dangerous if the bow submerges in the waves, a phenomenon called pounding. Sometimes, ships must change course or speed to stop violent rolling or pitching.

How it has been convincingly shown in scientific studies of the 21st century, controllability of some vessels decreases dramatically in some cases that are conditioned by effects of the bifurcation memory. This class of vessels includes ships with high manoeuvring capabilities, aircraft and controlled underwater vehicles designed to be unstable in steady-state motion that are interesting in terms of applications. These features must be considered in designing ships and in their control in critical situations.

## LIFECYCLE

A ship will pass through several stages during its career. The first is usually an initial contract to build the ship, the details of which can vary widely based on relationships between the shipowners, operators, designers and the shipyard. Then, the design phase carried out by a naval architect. Then the ship is constructed in a shipyard. After construction, the vessel is launched and goes into service. Ships end their careers in a number of ways, ranging from shipwrecks to service as a museum ship to the scrapyard.

### DESIGN

A vessel's design starts with a specification, which a naval architect uses to create a project outline, assess required dimensions, and create a basic layout of spaces and a rough displacement. After this initial rough draft, the architect can create an initial hull design, a general profile and an initial overview of the ship's propulsion. At this stage, the designer can iterate on the ship's design, adding detail and refining the design at each stage.

The designer will typically produce an overall plan, a general specification describing the peculiarities of the vessel, and construction blueprints to be used at the building site. Designs for larger or more complex vessels may also include sail plans, electrical schematics, and plumbing and ventilation plans.

As environmental laws are strictening, ship designers need to create their design in such a way that the ship -when it nears its end-of-term- can be disassmbled or disposed easily and that waste is reduced to a minimum.

### CONSTRUCTION

Ship construction takes place in a shipyard, and can last from a few months for a unit produced in series, to several years to reconstruct a wooden boat like the frigate *Hermione*, to more than 10 years for an aircraft carrier. Hull materials and vessel size play a large part in determining the method of construction. The hull of a mass-produced fiberglass sailboat is constructed from a mold, while the steel hull of a cargo ship is made from large sections welded together as they are built.

Generally, construction starts with the hull, and on vessels over about 30 meters (98 ft), by the laying of the keel. This is done in a drydock or on land. Once the hull is assembled and painted, it is launched. The last stages, such as raising the superstructure and adding equipment and accommodation, can be done after the vessel is afloat.

Once completed, the vessel is delivered to the customer. Ship launching is often a ceremony of some significance, and is usually when the vessel is formally named.

A typical small rowboat can cost under US$100, $1,000 for a small speedboat, tens of thousands of dollars for a cruising sailboat, and about $2,000,000 for a Vendée Globe class sailboat. A 25 meters (82 ft) trawler may cost $2.5 million, and a 1,000-person-capacity high-speed passenger ferry can cost in the neighborhood of $50 million. A ship's cost partly depends on its complexity: a small, general cargo ship will cost $20 million, a Panamax-sized bulk carrier around $35 million, a supertanker around $105 million and a large LNG carrier nearly $200 million. The most expensive ships generally are so because of the cost of embedded electronics: a *Seawolf*-class submarine costs around $2 billion, and an aircraft carrier goes for about $3.5 billion.

## REPAIR AND CONVERSION

Ships undergo nearly constant maintenance during their career, whether they be underway, pierside, or in some cases, in periods of reduced operating status between charters or shipping seasons.

Most ships, however, require trips to special facilities such as a drydock at regular intervals. Tasks often done at drydock include removing biological growths on the hull, sandblasting and repainting the hull, and replacing sacrificial anodes used to protect submerged equipment from corrosion. Major repairs to the propulsion and steering systems as well as major electrical systems are also often performed at dry dock.

Vessels that sustain major damage at sea may be repaired at a facility equipped for major repairs, such as a shipyard. Ships may also be converted for a new purpose: oil tankers are often converted into floating production storage and offloading units.

## END OF SERVICE

Most ocean-going cargo ships have a life expectancy of between 20 and 30 years. A sailboat made of plywood or fiberglass can last between 30 and 40 years. Solid wooden ships can last much longer but require regular maintenance. Carefully maintained steel-hulled yachts can have a lifespan of over 100 years.

As ships age, forces such as corrosion, osmosis, and rotting compromise hull strength, and a vessel becomes too dangerous to sail. At this point, it can be scuttled at sea or scrapped by shipbreakers. Ships can also be used as museum ships, or expended to construct breakwaters or artificial reefs.

Many ships do not make it to the scrapyard, and are lost in fires, collisions, grounding, or sinking at sea. There are more than 3 million shipwrecks on the ocean floor, the United Nations estimates. The Allies lost some 5,150 ships during World War II.

## MEASURING SHIPS

One can measure ships in terms of overall length, length of the ship at the waterline, beam (breadth), depth (distance between the crown of the weather deck and the top of the keelson), draft (distance between the highest waterline and the bottom of the ship) and tonnage. A number of different tonnage definitions exist and are used when describing merchant ships for the purpose of tolls, taxation, etc.

In Britain until Samuel Plimsoll's Merchant Shipping Act of 1876, ship-owners could load their vessels until their decks were almost awash, resulting in a dangerously unstable condition. Anyone who signed on to such a ship for a voyage and, upon realizing the danger, chose to leave the ship, could end up in jail.

Plimsoll, a Member of Parliament, realised the problem and engaged some engineers to derive a fairly simple formula to determine the position of a line on the side of any specific ship's hull which, when it reached the surface of the water during loading of cargo, meant the ship had reached its maximum safe loading level. To this day, that mark, called the "Plimsoll Line", exists on ships' sides, and consists of a circle with a horizontal line through the centre. On the Great Lakes of North America the circle is replaced with a diamond.

Because different types of water (summer, fresh, tropical fresh, winter north Atlantic) have different densities, subsequent regulations required painting a group of lines forward of the Plimsoll mark to indicate the safe depth (or freeboard above the surface) to which a specific ship could load in water of various densities. Hence the "ladder" of lines seen forward of the Plimsoll mark to this day. This is called the "freeboard mark" or "load line mark" in the marine industry.

## SHIP POLLUTION

Ship pollution is the pollution of air and water by shipping. It is a problem that has been accelerating as trade has become increasingly globalized, posing an increasing threat to the world's oceans and waterways as globalization continues. It is expected that, "...shipping traffic to and from the USA is projected to double by 2020." Because of increased traffic in ocean ports, pollution from ships also directly affects coastal areas. The pollution produced affects biodiversity, climate, food, and human health. However, the degree to which humans are polluting and how it affects the world is highly debated and has been a hot international topic for the past 30 years.

## OIL SPILLS

Oil spills have devastating effects on the environment. Crude oil contains polycyclic aromatic hydrocarbons (PAHs) which are very difficult to clean up, and last for years in the sediment and marine environment. Marine species constantly exposed to PAHs can exhibit developmental problems, susceptibility to disease, and abnormal reproductive cycles.

By the sheer amount of oil carried, modern oil tankers must be considered something of a threat to the environment. An oil tanker can carry 2 million barrels (318,000 m) of crude oil, or 84,000,000 US gallons (69,940,000 imp gal; 318,000,000 L). This is more than six times the amount spilled in the widely known *Exxon Valdez* incident. In this spill, the ship ran aground and dumped 10,800,000 US gallons (8,993,000 imp gal; 40,880,000 L) of oil into the ocean in March 1989. Despite efforts of scientists, managers, and volunteers, over 400,000 seabirds, about 1,000 sea otters, and immense numbers of fish were killed.

The International Tanker Owners Pollution Federation has researched 9,351 accidental spills since 1974. According to this study, most spills result from routine operations such as loading cargo, discharging cargo, and taking on fuel oil. 91% of the operational oil spills were small, resulting in less than 7 tons per spill. Spills resulting from accidents like collisions, groundings, hull failures, and explosions are much larger, with 84% of these involving losses of over 700 tons.

Following the *Exxon Valdez* spill, the United States passed the Oil Pollution Act of 1990 (OPA-90), which included a stipulation that all tankers entering its waters be double-hulled by 2015. Following the sinkings of the *Erika* (1999) and *Prestige* (2002), the European Union passed its own stringent anti-pollution packages (known as Erika I, II, and III), which require all tankers entering its waters to be double-hulled by 2010. The Erika packages are controversial because they introduced the new legal concept of "serious negligence".

## BALLAST WATER

When a large vessel such as a container ship or an oil tanker unloads cargo, seawater is pumped into other compartments in the hull to help stabilize and balance the ship. During loading, this ballast water is pumped out from these compartments.

One of the problems with ballast water transfer is the transport of harmful organisms. Meinesz believes that one of the worst cases of a single invasive species causing harm to an ecosystem can be attributed to a seemingly harmless jellyfish. *Mnemiopsis leidyi,* a species of comb jellyfish that inhabits estuaries from the United States to the Valdés peninsula in Argentina along the Atlantic coast, has caused notable damage in the Black Sea. It was first introduced in 1982, and thought to have been transported to the Black Sea in a ship's ballast water.

The population of the jellyfish shot up exponentially and, by 1988, it was wreaking havoc upon the local fishing industry. "The anchovy catch fell from 204,000 tonnes (225,000 short tons; 201,000 long tons) in 1984 to 200 tonnes (220 short tons; 197 long tons) in 1993; sprat from 24,600 tonnes (27,100 short tons; 24,200 long tons) in 1984 to 12,000 tonnes (13,200 short tons; 11,800 long tons) in 1993; horse mackerel from 4,000 tonnes (4,410 short tons; 3,940 long tons) in 1984 to zero in 1993."

Now that the jellyfish have exhausted the zooplankton, including fish larvae, their numbers have fallen dramatically, yet they continue to maintain a stranglehold on the ecosystem. Recently the jellyfish have been discovered in the Caspian Sea. Invasive species can take over once occupied areas, facilitate the spread of new diseases, introduce new genetic material, alter landscapes and jeopardize the ability of native species to obtain food. "On land and in the sea, invasive species are responsible for about 137 billion dollars in lost revenue and management costs in the U.S. each year."

Ballast and bilge discharge from ships can also spread human pathogens and other harmful diseases and toxins potentially causing health issues for humans and marine life alike. Discharges into coastal waters, along with other sources of marine pollution, have the potential to be toxic to marine plants, animals, and microorganisms, causing alterations such as changes in growth, disruption of hormone cycles, birth defects, suppression of the immune system, and disorders resulting in cancer, tumors, and genetic abnormalities or even death.

## EXHAUST EMISSIONS

Exhaust emissions from ships are considered to be a significant source of air pollution. "Seagoing vessels are responsible for an estimated 14 percent of emissions of nitrogen from fossil fuels and 16 percent of the emissions of sulfur from petroleum uses into the atmosphere." In Europe ships make up a large percentage of the sulfur introduced to the air, "...as much sulfur as all the cars, lorries and factories in Europe put together." "By 2010, up to 40% of air pollution over land could come from ships." Sulfur in the air creates acid rain which damages crops and buildings. When inhaled sulfur is known to cause respiratory problems and increase the risk of a heart attack.

## SHIP BREAKING

Ship breaking or ship demolition is a type of ship disposal involving the breaking up of ships for scrap recycling, with the hulls being discarded in ship graveyards. Most ships have a lifespan of a few decades before there is so much wear that refitting and repair becomes uneconomical. Ship breaking allows materials from the ship, especially steel, to be reused.

In addition to steel and other useful materials, however, ships (particularly older vessels) can contain many substances that are banned or considered dangerous in developed countries. Asbestos and polychlorinated biphenyls (PCBs) are typical examples. Asbestos was used heavily in ship construction until it was finally banned in most of the developed world in the mid 1980s.

Currently, the costs associated with removing asbestos, along with the potentially expensive insurance and health risks, have meant that ship-breaking in most developed countries is no longer economically viable. Removing the metal for scrap can potentially cost more than the scrap value of the metal itself.

In the developing world, however, shipyards can operate without the risk of personal injury lawsuits or workers' health claims, meaning many of these shipyards may operate with high health risks. Protective equipment is sometimes absent or inadequate. Dangerous vapors and fumes from burning materials can be inhaled, and dusty asbestos-laden areas around such breakdown locations are commonplace.

Aside from the health of the yard workers, in recent years, ship breaking has also become an issue of major environmental concern. Many developing nations, in which ship breaking yards are located, have lax or no environmental law, enabling large quantities of highly toxic materials to escape into the environment and causing serious health problems among ship breakers, the local population and wildlife. Environmental campaign groups such as Greenpeace have made the issue a high priority for their campaigns.

## BUOYANCY

A floating boat displaces its weight in water. The material of the boat hull may be denser than water, but if this is the case then it forms only the outer layer. If the boat floats, the mass of the boat (plus contents) *as a whole* divided by the volume *below the waterline* is equal to the density of water (1 kg/l). If weight is added to the boat, the volume below the waterline will increase to keep the weight balance equal, and so the boat sinks a little to compensate.

## SHIPBUILDING

Shipbuilding is the construction of ships and floating vessels. It normally takes place in a specialized facility known as a shipyard. Shipbuilders, also called shipwrights, follow a specialized occupation that traces its roots to before recorded history. Shipbuilding and ship repairs, both commercial and military, are referred to as "naval engineering". The construction of boats is a similar activity called boat building. The dismantling of ships is called ship breaking.

## HISTORY

### Prehistory

Archaeological evidence indicates that humans arrived on Borneo at least 120,000 years ago, probably by sea from the Asian mainland during an ice age period when the sea was lower and distances between islands shorter. The ancestors of Australian Aborigines and New Guineans also went across the Lombok Strait to Sahul by boat over 50,000 years ago.

### 4th Millennium BC

Evidence from Ancient Egypt shows that the early Egyptians knew how to assemble planks of wood into a ship hull as early as 3000 BC. The Archaeological Institute of America reports that some of the oldest ships yet unearthed are known as the Abydos boats. These are a group of 14 discovered ships in Abydos that were constructed of wooden planks which were "sewn" together.

Discovered by Egyptologist David O'Connor of New York University, woven straps were found to have been used to lash the planks together, and reeds or grass stuffed between the planks helped to seal the seams. Because the ships are all buried together and near a mortuary belonging to Pharaoh Khasekhemwy, originally they were all thought to have belonged to him, but one of the 14 ships dates to 3000 BC, and the associated pottery jars buried with the vessels also suggest earlier dating. The ship dating to 3000 BC was about 25 m, 75 feet long and is now thought to perhaps have belonged to an earlier pharaoh. According to professor O'Connor, the 5,000-year-old ship may have even belonged to Pharaoh Aha.

### 3rd Millennium BC

Early Egyptians also knew how to assemble planks of wood with treenails to fasten them together, using pitch for caulking the seams. The "Khufu ship", a 43.6-meter vessel sealed into a pit in the Giza pyramid complex at the foot of the Great Pyramid of Giza in the Fourth Dynasty around 2500 BC, is a full-size surviving example which may have fulfilled the symbolic function of a solar barque. Early Egyptians also knew how to fasten the planks of this ship together with mortise and tenon joints.

The oldest known tidal dock in the world was built around 2500 BC during the Harappan civilisation at Lothal near the present day Mangrol harbour on the Gujarat coast in India. Other ports were probably at Balakot and Dwarka. However, it is probable that many small-scale ports, and not massive ports, were used for the Harappan maritime trade. Ships from the harbour at these ancient port cities established trade with Mesopotamia. Shipbuilding and boatmaking may have been prosperous industries in ancient India. Native labourers may have manufactured the flotilla of boats used by Alexander the Great to navigate across the Hydaspes and even the Indus, under Nearchos. The Indians also exported teak for shipbuilding to ancient Persia. Other references to Indian timber used for shipbuilding is noted in the works of Ibn Jubayr.

### 2nd Millennium BC

The ships of Ancient Egypt's Eighteenth Dynasty were typically about 25 meters (80 ft) in length, and had a single mast, sometimes consisting of two poles lashed together at the top making an "A" shape. They mounted a single square sail on a yard, with an additional spar along the bottom of the sail. These ships could also be oar propelled. The ocean and sea going ships of Ancient Egypt were constructed with cedar wood, most likely hailing from Lebanon.

The ships of Phoenicia seem to have been of a similar design. The Greeks and probably others introduced the use of multiple banks of oars for additional speed, and the ships were of a light construction for speed and so they could be carried ashore.

### 1st Millennium BC

The naval history of China stems back to the Spring and Autumn Period (722 BC–481 BC) of the ancient Chinese Zhou Dynasty. The Chinese built large rectangular barges known as "castle ships", which were essentially floating fortresses complete with multiple decks with guarded ramparts.

### Early 1st Millennium AD

The ancient Chinese also built ramming vessels as in the Greco-Roman tradition of the trireme, although oar-steered ships in China lost favor very early on since it was in the 1st century China that the stern-mounted rudder was first developed. This was dually met with the introduction of the Han Dynasty junk ship design in the same century.

Archeological investigations done at Portus near Rome have revealed inscriptions indicating the existence of a 'guild of shipbuilders' during the time of Hadrian.

### Medieval Europe, Song China, Abbasid Caliphate, Pacific Islanders

Viking longships developed from an alternate tradition of clinker-built hulls fastened with leather thongs. Sometime around the 12th century, northern European ships began to be built with a straight sternpost, enabling the mounting of a rudder, which was much more durable than a steering oar held over the side. Development in the Middle Ages favored "round ships", with a broad beam and heavily curved at both ends. Another important ship type was the galley which was constructed with both sails and oars.

An insight into ship building in the North Sea/Baltic areas of the early medieval period was found at Sutton Hoo, England, where a ship was buried with a chieftain. the ship was 26 metres (85 ft) long and, 4.3 metres (14 ft) wide. Upward from the keel, the hull was made by overlapping nine planks on either side with rivets fastening the oaken planks together. It could hold upwards of thirty men.

The first extant treatise on shipbuilding was written ca. 1436 by Michael of Rhodes, a man who began his career as an oarsman on a Venetian galley in 1401 and worked his way up into officer positions. He wrote and illustrated a book that contains a treatise on ship building, a treatise on mathematics, much material on astrology, and other materials. His treatise on shipbuilding treats three kinds of galleys and two kinds of round ships.

Outside Medieval Europe, great advances were being made in shipbuilding. The shipbuilding industry in Imperial China reached its height during the Song Dynasty, Yuan Dynasty, and early Ming Dynasty, building commercial vessels that by the end of this period were to reach a size and sophistication far exceeding that of contemporary Europe.

The mainstay of China's merchant and naval fleets was the junk, which had existed for centuries, but it was at this time that the large ships based on this design were built. During the Sung period (960–1279 AD), the establishment of China's first official standing navy in 1132 AD and the enormous increase in maritime trade abroad (from Heian Japan to Fatimid Egypt) allowed the shipbuilding industry in provinces like Fujian to thrive as never before. The largest seaports in the world were in China and included Guangzhou, Quanzhou, and Xiamen.

In the Islamic world, shipbuilding thrived at Basra and Alexandria, the dhow, felucca, baghlah and the sambuk, became symbols of successful maritime trade around the Indian Ocean; from the ports of East Africa to Southeast Asia and the ports of Sindh and Hind (India) during the Abbasid period.

At this time islands spread over vast distances across the Pacific Ocean were being colonised by the Melenesians and Polynesians, who built giant canoes and progressed to great catamarans.

## EARLY MODERN

With the development of the carrack, the west moved into a new era of ship construction by building the first regular ocean going vessels. In a relatively short time, these ships grew to an unprecedented size, complexity and cost. Shipyards became large industrial complexes and the ships built were financed by consortia of investors These considerations led to the documentation of design and construction practices in what had previously been a secretive trade run by master shipwrights, and ultimately led to the field of naval architecture, where professional designers and draughtsmen played an increasingly important role.

Even so, construction techniques changed only very gradually. The ships of the Napoleonic Wars were still built more or less to the same basic plan as those of the Spanish Armada of two centuries earlier but there had been numerous subtle improvements in ship design and construction throughout this period. For instance, the introduction of tumblehome; adjustments to the shapes of sails and hulls; the introduction of the wheel; the introduction of hardened copper fastenings below the waterline; the introduction of copper sheathing as a deterrent to shipworm and fouling; etc.

## INDUSTRIAL REVOLUTION

The industrial revolution made possible the use of new materials and designs that radically altered shipbuilding. Other than its widespread use in fastenings, Iron was gradually adopted in ship construction, initially in discrete areas in a wooden hull needing greater strength, (e.g. as deck knees, hanging knees, knee riders and the like).

Then, in the form of plates rivetted together and made watertight, it was used to form the hull itself. Initially copying wooden construction traditions with a frame over which the hull was fastened, Isambard Kingdom Brunel's *Great Britain* of 1843 was the first radical new design, being built entirely of wrought iron. Despite her success, and the great savings in cost and space provided by the iron hull, compared to a copper sheathed counterpart, there remained problems with fouling due to the adherence of weeds and barnacles. As a result composite construction remained the dominant approach where fast ships were required, with wooden timbers laid over an iron frame (the *Cutty Sark* is a famous example).

Later *Great Britain's* iron hull was sheathed in wood to enable it to carry a copper-based sheathing. Brunel's *Great Eastern* represented the next great development in shipbuilding. Built in association with John Scott Russell, it used longitudinal stringers for strength, inner and outer hulls, and bulkheads to form multiple watertight compartments. Steel also supplanted wrought iron when it became readily available in the latter half of the 19th century, providing great savings when compared with iron in cost and weight. Wood continued to be favored for the decks.

## WORLDWIDE SHIPBUILDING INDUSTRY

After the Second World War, shipbuilding (which encompasses the shipyards, the marine equipment manufacturers, and many related service and knowledge providers) grew as an important and strategic industry in a number of countries around the world. This importance stems from:

- The large number of skilled workers required directly by the shipyard, along with supporting industries such as steel mills, railroads and engine manufacturers; and
- A nation's need to manufacture and repair its own navy and vessels that support its primary industries

Historically, the industry has suffered from the absence of global rules and a tendency towards (state-supported) over-investment due to the fact that shipyards offer a wide range of technologies, employ a significant number of workers, and generate income as the shipbuilding market is global.

Shipbuilding is therefore an attractive industry for developing nations. Japan used shipbuilding in the 1950s and 1960s to rebuild its industrial structure; South Korea started to make shipbuilding a strategic industry in the 1970s, and China is now in the process of repeating these models with large state-supported investments in this industry. Conversely, Croatia is privatising its shipbuilding industry.

As a result, the world shipbuilding market suffers from over-capacities, depressed prices (although the industry experienced a price increase in the period 2003–2005 due to strong demand for new ships which was in excess of actual cost increases), low profit margins, trade distortions and widespread subsidisation. All efforts to address the problems in the OECD have so far failed, with the 1994 international shipbuilding agreement never entering into force and the 2003–2005 round of negotiations being paused in September 2005 after no agreement was possible.

After numerous efforts to restart the negotiations these were formally terminated in December 2010. The OECD's Council Working Party on Shipbuilding (WP6) will continue its efforts to identify and progressively reduce factors that distort the shipbuilding market.

Where state subsidies have been removed and domestic industrial policies do not provide support, in high-cost nations shipbuilding has usually gone into steady, if not rapid, decline. The British shipbuilding industry is a prime example of this. From a position in the early 1970s where British yards could still build the largest types of sophisticated merchant ships, British shipbuilders today have been reduced to a handful specialising in defence contracts and repair work.

In the U.S.A., the Jones Act (which places restrictions on the ships that can be used for moving domestic cargoes) has meant that merchant shipbuilding has continued, but such protection has failed to penalise shipbuilding inefficiencies. The consequence of this is contract prices that are far higher than those of any other nation building oceangoing ships.

**Present day Shipbuilding**

China is an emerging shipbuilder overtook South Korea during the 2008-2010 global financial crisis as they won new orders for medium and small-sized container ships. China is now firmly the world's largest shipbuilder with 45% of the world's total orders, and its quality and technology have improved very much.

Today, South Korea is the world's second largest shipbuilding country with a global market share of 29% in 2012. South Korea leads in the production of large vessels such as cruise liners, super tankers, LNG carriers, drill ships, and large container ships. In the 3rd quarter of 2011, South Korea won all 18 orders for LNG carriers, 3 out of 5 drill ships and 5 out of 7 large container ships.

Japan had been the dominant ship building country from the 1960s through to the end of 1990s but gradually lost its competitive advantage to the emerging industry in South Korea which had the advantages of much cheaper wages, strong government backing and a cheaper currency. South Korean production overtook Japan's in 2003 and Japanese market share has since fallen sharply.

The market share of European ship builders began to decline in the 1960s as they lost work to the Japanese in the same way as Japanese builders have lost work to South Koreans more recently; Europe's production is now a tenth of South Korea's and is primarily military, although cruise liners and some cargo ships are still built in Italy, Finland, France, Germany and Denmark. The output of the United States also underwent a similar change.

South Korea's shipyards are highly efficient, with the world's largest shipyard in Ulsan operated by Hyundai Heavy Industries slipping a newly built, $80 million vessel into the water every four working days. South Korea's "big three" shipbuilders, Hyundai Heavy Industries, Samsung Heavy Industries, and Daewoo Shipbuilding & Marine Engineering, dominate global shipbuilding, with STX Shipbuilding, Hyundai Samho Heavy Industries, Hanjin Heavy Industries, and Sungdong Shipbuilding & Marine Engineering also ranking among the top ten shipbuilders in the world. In 2007, STX Shipbuilding further strengthened South Korea's leading position in the industry by acquiring Aker Yards, the largest shipbuilding group in Europe. (The former Aker Yards was renamed STX Europe in 2008). In the first half of 2011, South Korean shipbuilders won new orders to build 25 LNG carriers, out of the total 29 orders placed worldwide during the period.

**Modern shipbuilding manufacturing techniques**

Modern shipbuilding makes considerable use of prefabricated sections. Entire multi-deck segments of the hull or superstructure will be built elsewhere in the yard, transported to the building dock or slipway, then lifted into place. This is known as "block construction".

The most modern shipyards pre-install equipment, pipes, electrical cables, and any other components within the blocks, to minimize the effort needed to assemble or install components deep within the hull once it is welded together. Ship design work, also called naval architecture, may be conducted using a ship model basin. Modern ships, since roughly 1940, have been produced almost exclusively of welded steel. Early welded steel ships used steels with inadequate fracture toughness, which resulted in some ships suffering catastrophic brittle fracture structural cracks.

Since roughly 1950, specialized steels such as ABS Steels with good properties for ship construction have been used. Although it is commonly accepted that modern steel has eliminated brittle fracture in ships, some controversy still exists. Brittle fracture of modern vessels continues to occur from time to time because grade A and grade B steel of unknown toughness or fracture appearance transition temperature (FATT) in ships' side shells can be less than adequate for all ambient conditions.

**Ship Repair Industry**

All ships need maintenance and repairs. A part of these jobs must be carried out under the supervision of the Classification Society. A lot of maintenance is carried out while at sea or in port by ship's staff. However a large number of repair and maintenance works can only be carried out while the ship is out of commercial operation, in a Shiprepair Yard. Prior to undergoing repairs, tankers must dock at a Deballasting Station for completing the tank cleaning operations and pumping ashore its slops (dirty cleaning water and hydrocarbon residues).

# 2

# Ship's Equipment

## SHIPBOARD FIRE-FIGHTING EQUIPMENT

### GENERAL

The requirements for tankers' fire-fighting equipment are laid down by the regulations of the particular country in which the tanker is registered. The theory of fire-fighting and the types of fire that may be encountered are discussed in Chapter 5.

### TANKER FIXED FIRE-FIGHTING INSTALLATIONS - COOLING

All tankers are provided with a water fire-fighting system consisting of pumps with a permanent underwater connection, a fire-main with hydrant points, fire hoses complete with couplings, and jet nozzles or, preferably, jet/spray nozzles. A sufficient number of hydrants are provided and located so as to ensure that two jets of water can reach any part of the tanker. In cold weather, the freezing of fire-mains and hydrants should be prevented by continuously bleeding water overboard from hydrants at the extreme end of each fire-main. Alternatively, all low points of the fire-main may be kept drained.

### TANKER FIXED FIRE-FIGHTING INSTALLATIONS - SMOTHERING

One or more of the different smothering systems listed below may be installed on board tankers.

#### Carbon Dioxide Flooding System

This system is designed to fight fires in the engine room, boiler room and pumproom. The system normally consists of a battery of large carbon dioxide cylinders. The carbon dioxide is piped from the cylinder manifold to suitable points having diffusing nozzles. An alarm should be activated in the compartment before the carbon dioxide is released to give personnel time to evacuate the compartment.

### Foam Systems

Foam systems are used for fighting fire in the cargo spaces, on the cargo deck, in the pumproom or in the engine spaces. A foam system has storage tanks containing foam concentrate. Water from the fire pumps picks up the correct proportion of foam concentrate from the tank through a proportioner and the foam solution is then conveyed through permanent supply lines to offtake points, fixed foam monitors or, in the case of engine room installations, to fixed dispersal nozzles.

### Water Fog

A water fog system consists of high pressure water lines and special fog nozzles. A ring of nozzles around the inside of the tank opening effectively blankets a cargo tank hatch fire. Some tankers are also fitted with fixed pressurised water fog systems for protecting specific parts of the engine room, such as oil fuel treatment spaces, boiler firing platforms, small machinery spaces and pumprooms.

### Water Curtain

Some tankers have a fixed system to give a protective water curtain between the cargo deck and the superstructure.

### Inert Gas System

The purpose of an inert gas system is to prevent cargo tank fires or explosions. It is not a fixed fire-fighting installation but, in the event of a fire, the system may be of assistance in controlling the fire and preventing explosions.

## PORTABLE FIRE EXTINGUISHERS

All tankers are provided with a range of portable fire extinguishers to meet the requirements of the respective legislation. All fire extinguishers should at all times be in good order and available for immediate use.. As a minimum, all fire extinguishers should be formally checked for proper location, charging pressure and condition annually. Consideration should be given to providing portable extinguishers, suitable for use on Class A fires, and dedicated to deployment at the tanker's manifold when in port.

### Types of Portable Fire Extinguisher

In addition to fire hose reels for water extinguishing of Class A type fires involving combustible materials, such as wood, paper and fabrics, all tankers are provided with a range of portable fire extinguishers. Table provides an overview of the types of extinguisher likely to be found on board a tanker and their uses. Class D type fires are included mainly for completeness.

**Portable fire extinguishing media and their uses**

| Type of Fire | Class A | Class B | Class C | Class D | Class F | |
|---|---|---|---|---|---|---|
| Fire Extinguishing Medium | Fires Involving Solid Materials (e.g. wood, paper, fabrics) | Fires Involving Liquids or Liquefiable Solids | Fires Involving Gases | Fires Involving Metals (e.g. magne-sium, titanium, potassium and sodium) | Fires Involving Cooking Media in Cooking Ap-pliances | Electrical Equipment Fires |
| Water/Hose Reels | ✓ | | | | | |
| Water with Additive | ✓ | | | | | |
| Spray Foam | ✓ | ✓ | | | | |
| Dry Chemical | ✓ | ✓ | ✓ | | | ✓ |
| $CO_2$ Gas | | ✓ | | | | ✓ |
| Wet Chemical | ✓ | | | | ✓ | |
| Fire Blanket | | | | | ✓ | |
| Designed to match a particular type of fire | | | | ✓ | | |

## GAS TESTING EQUIPMENT

The safe management of operations on board tankers is often dependent upon the crew's ability to determine the composition of the ambient atmosphere or the atmosphere in an enclosed space. Tanker crews need to measure the oxygen, flammable and toxic gas concentrations in an atmosphere. This will enable them to detect the presence of any explosive mixtures, toxic vapours or oxygen deficiency that may present a risk of explosion or hazard to personnel.

On tankers fitted with an inert gas system, there is the additional need to measure the oxygen content of inert gas as part of the safe management of cargo tank atmospheres.

### SUMMARY OF GAS TESTING TASKS

#### Atmosphere Monitoring

The external atmosphere should be monitored for:

- Flammable vapour when undertaking Hot Work. This is achieved by using a flammable gas indicator, capable of measuring gas to the Lower Explosive Limit (LEL) and with the scale graduated as a percentage of this limit.

- Toxic vapours when loading cargoes containing toxic components and when undertaking gas freeing operations following the carriage of such cargoes. This is achieved by using an instrument capable of measuring concentrations of toxic gases in the human toxicity range, usually calibrated in parts per million.

### Enclosed Space Monitoring

Prior to permitting entry into an enclosed space, measurements must be taken to detect the presence of hydrocarbon gas, to confirm normal oxygen levels and, if applicable, to detect the presence of any toxic vapours.

Measurement to ensure that the atmosphere is free of harmful hydrocarbon vapour is undertaken using a flammable gas indicator capable of measuring gas to the Lower Explosive Limit (LEL) and with the scale graduated as a percentage of this limit (% LEL).

An oxygen analyser is used to determine that the normal level of oxygen in air of 20.9% by volume is present. Where toxic vapour may be present in the space to be entered, the atmosphere should also be tested with an instrument capable of measuring concentrations of toxic gases in the human toxicity range, usually calibrated in parts per million.

### Inert Gas Atmosphere Management

Tankers fitted with an inert gas system should be equipped with an oxygen analyser for determining the quality of the inert gas and for measuring the levels of oxygen in the cargo tanks. A gas indicator capable of measuring the percentage of flammable gas by volume (% Vol) in an inerted atmosphere is also required for safe management of operations that include the purging and gas freeing of cargo tanks.

## THE PROVISION OF GAS MEASURING INSTRUMENTS

It is recommended that a tanker carrying cargoes that are likely to emit a toxic or flammable gas, or to cause oxygen depletion in a cargo space, be provided with an appropriate instrument for measuring the concentration of gas or oxygen in the air, together with detailed instructions for its use. Implicit in the above recommendation is the requirement that the tanker operator provides the correct instrument for each gas test required.

It should be noted that the different gas testing functions can be incorporated into a multi-function gas measuring instrument. The gas measurement instrumentation on board a tanker should form a comprehensive and integrated system that addresses all the necessary applications identified by the operator.

The instruments should be fit for the task to which they are applied and users should be made aware of the particular applications and limitations of each instrument. Users of gas measuring instruments should be trained in the proper use of the equipment, to a level suited to their work duties.

## ALARM FUNCTIONS ON GAS MEASURING INSTRUMENTS

Alarms should only be fitted to instruments that are to be used where an audible warning is necessary, such as a personal gas alarm monitor. Analytical instruments that are used to provide numerical values for gases and vapours for dangerous space entry certification do not need to have an alarm function. Instruments with an alarm capability should be designed so that the alarm inhibit and activate function cannot be changed by the instrument operator.

This is to avoid the possibility of inappropriate or accidental inhibition of the alarm function. The use of different instruments for testing atmospheres for entry certification, and for monitoring atmospheres with a personal monitor during the entry operation, reduces the probability of an accident due to an instrument malfunction.

It is therefore recommended that the testing instrument is not also used as the personal alarm instrument during the entry operation.

## SAMPLING LINES

If fitted, sampling lines should be suitable for the intended service and be impervious to the gases present in the atmospheres being monitored. They should also be resistant to the effects of hot wash water.

## CALIBRATION

Calibration should not be confused with operational testing. The accuracy of measurement equipment should be in accordance with the manufacturer's stated standards. Equipment should, on initial supply, have a calibration certificate, traceable where possible to internationally recognised standards. Thereafter, procedures for management of the calibration certification process should form part of the on board Safety Management System.

These procedures may include on board calibration in line with the manufacturer's guidelines and/or equipment being periodically landed to a recognised testing facility for calibration, either on a timed basis, or during the tanker's refit, or when the accuracy of the equipment is considered to be outside the manufacturer's stated accuracy. Calibration certificates, showing the instrument's serial number, the calibration date and the calibration gas or the method of calibration used, together with reference to applicable standards, should be provided for retention on board.

Instruments are typically calibrated using a calibration gas consistent with the use of the instrument, such as propane or butane. The calibration gas used should be marked on the instrument.

The use of an inappropriate gas for calibration could result in erroneous readings during operation, even though the instrument appears to be operating correctly. Instruments should only be dismantled by persons who are qualified and certified to carry out such work.

## OPERATIONAL TESTING AND INSPECTION

Gas measuring instruments should be tested in accordance with the manufacturer's instructions before the commencement of operations requiring their use. Such tests are designed only to ensure that the instrument is working properly. They should not be confused with calibration. Instruments should only be used if the tests indicate that the instrument is giving accurate readings and that alarms, if fitted, are operating at the pre-determined set points. Physical checks should include (if applicable):

- Hand pump.
- Extension tubes.
- Tightness of connections.
- Batteries.
- Housing and case.

Instruments not passing these operational tests should be re-calibrated before they are returned to operational use. If this is not possible, they should be removed from service and clearly labelled to denote that they are not to be used. During operations, it is important to check the instrument and sample lines for leakage occasionally, since the ingress of air will dilute the sample and give false readings.

Leak testing may be carried out by pinching the end of the sample line and squeezing the aspirator bulb. The bulb should not expand as long as the sample line is pinched. During extended operations, the tanker operator should determine the frequency at which operational checks should be made. The results of the tests and inspections should be recorded. These procedures should be documented in the Safety Management System.

## DISPOSABLE PERSONAL GAS MONITORS

Disposable personal gas monitors should be periodically tested in accordance with the manufacturer's recommendations to confirm that they are operating correctly.

Disposable gas detection monitors, which cannot be re-calibrated, should be safely disposed of when the calibration expiry date is reached. For this reason, it is important to record the date when disposable instruments are first commissioned in order to establish their expiry date.

# LIFTING EQUIPMENT

## INSPECTION AND MAINTENANCE

All shipboard lifting equipment, such as is used for the handling of cargo transfer equipment and/or gangways, should be examined at intervals not exceeding one year and load tested at least every 5 years unless local, national or company regulations require more frequent examinations. Lifting equipment includes:

- Cargo hose handling cranes, derricks, davits and gantries.
- Gangways and associated cranes and davits.
- Store cranes and davits.
- Chain blocks, hand winches and similar mechanical devices.
- Personnel lifts and hoists.
- Strops, slings, chains and other ancillary equipment.

All equipment should be tested by suitably qualified individuals or authorities and be clearly marked with its Safe Working Load (SWL), serial number and test date.

The tanker should ensure that all maintenance of lifting equipment is carried out in accordance with manufacturer's guidelines. Routine checks should be included within the tanker's planned maintenance system. All records of tests and inspections should be recorded in the ship's Lifting Equipment Register. These records should be available for inspection by Terminal Representatives when their personnel are involved in lifting operations using tanker's equipment.

**TRAINING**

Lifting equipment should only be operated by personnel who are trained and proven to be competent in its operation.

# 3

# Ship Construction

## YARD (SAILING)

A yard is a spar on a mast from which sails are set. It may be constructed of timber or steel or from more modern materials like aluminium or carbon fibre. Although some types of fore and aft rigs have yards, the term is usually used to describe the horizontal spars used on square rigged sails. In addition, for some decades after square sails were generally dispensed with, some yards were retained for deploying wireless (radio) aerials and signal flags.

## PARTS OF THE YARD

- Bunt: The short section of the yard between the *slings* that attach it to the mast.
- Quarters: The port and starboard quarters form the bulk of the yard, extending from the slings to the fittings for the lifts and braces.
- Yardarms: The outermost tips of the yard: outboard from the attachments for the lifts.

## CONTROLLING THE YARD

The yard can rotate around the mast to allow the direction of the vessel to be changed relative to the wind. When running directly downwind the yards are 'squared', pointing perpendicular to the ship's centre line. As the ship is steered closer to the wind the yards are braced round using the braces. When further rotation is obstructed by other bits of rigging (typically the shrouds), the yard is said to be braced "hard round" or "sharp up", as in "sharp up to port". This angle (normally about 45 degrees) limits how close to the wind a square rigged ship can sail.

The yards represent a considerable weight high above the vessel's centre of gravity. To increase stability, especially in heavy weather, some means is normally provided to lower some of the yards when they are not being used to set sails. In nineteenth-century warships (where a large crew was available) this was generally by physically "sending down" the upper yards from the masts and storing them on deck - along with, in many cases, the upper sections of the mast itself.

Merchant ships in the age of sail would also do this before sailing in the Southern Ocean. On modern tall ships the yards are not designed to be sent down on deck, but 'lifting yards' that can be raised and lowered along a short section of mast using a halyard are often used.

As well as rotating round the mast and moving up and down along it, the yards on many ships are designed to tilt relative to the mast. This allows the sails to be set more efficiently when the ship is heeled over by raising the leeward yardarm to bring the yard closer to the horizontal. This is achieved using the *lifts*, which run from each yardarm to the mast some way above. On some ships only the course lifts can be adjusted (the others being *fixed lifts* intended only to support the yard when not hoisted), with the influence of the course yard being sufficient to tilt all the sails. Some ships have their yards mounted on mechanical swivels with no possibility of tilting them.

**GOING ALOFT**

In order to set and stow the square sails, the crew must climb aloft and spread out along the yards. To do this, they stand in footropes suspended beneath the yard and balance themselves between that and the yard itself. The person working on the end of the yardarm has a separate footrope known as the flemish horse. *Jackstays* run along the top of the yard - the sail will be bent on to one of them, but a second one is often provided (particularly on larger yards) for the crew to hold onto. These are usually steel rods, but stiff cordage stretched between the yardarms was used in the past.

Almost all ships used in modern times are fitted with steel safety wires (sometimes erroneously called jackstays) along the yards to which sailors attach themselves using a harness. This is a relatively recent innovation - cargo-carrying and naval sailing ships were not so equipped and falling from the yard represented a real (though less than commonly imagined) risk.

**SETTING SQUARE SAILS**

The yard exists to allow square sails to be set to drive the ship. The top edge of the sail is 'bent on' (attached) to the yard semi-permanently. Clewlines and buntlines are led along the yard and from there to the mast and down to the deck. These allow the bottom of the sail to be hoisted up to the yard, so the sail is effectively folded in two.

In this state the sail is said to be "in its gear", that is ready for setting or stowing. To set the sail the clewlines and buntlines are let go, and the sheets (attached to the bottom corners - clews - of the sail) are adjusted to shape the sail to best catch the wind. A device called a "tye", a "tye chain", or a "tye rope" is used for lifting a given yard to the top of its travel to set the sail.

To stop using (i.e., to "hand") the sail, the sheets are released and the clewlines and buntlines are pulled tight. The sail folds in half - back in its gear - and so no longer catches the wind. Unless the sail is to be used again very soon, the next step is to "stow" it. To do this, the crew must go out along the yard in order to bundle the sail up tightly and tie it down with gaskets.

## MANNING THE YARDS

When coming into port, especially during the Tall Ships' Races many modern square riggers will 'man the yards'. All the crew not required on deck to handle the ship will go aloft and spread out along the yards. This manoeuvre was originally used to display the whole crew to the harbour authorities and the other ships present, to show that the ship's guns were not manned and hence her intentions were peaceful.

The Mexican Navy's training ship Cuauhtémoc is famous for manning its yards with its crew standing on the yards themselves, rather than in footropes.

## FORE AND AFT YARDS

As well as the square-rig yard described above, the traditional lateen rig is a triangular sail rigged fore and aft from a long yard mounted at an angle (downward sloping forward) from the mast. As well, some smaller fore and aft rigs use a yard. The spar at the head of a lugsail - a roughly square sail which is set fore-and-aft but requires more handling than a more modern gaff or Bermuda rig - is known as a yard, and probably developed from the original square-rig yard. The spar at the head of a gunter-rigged sail serves the function of a running topmast, but is not given that name. Some would call it a 'gaff', while others would use the name 'yard'.

## "SUN OVER THE YARDARM"

This phrase is widely used, both afloat and ashore, to indicate that the time of day has been reached at which it is acceptable, variously, to have lunch or (more commonly) to have an alcoholic beverage. In modern parlance, the latter usage typically refers to early evening, but the phrase is thought originally to have referred to late morning and to the sun's ascent past a particular yard. The actual time that the sun would pass a particular yard would depend greatly on the ship's latitude and heading, as well as the height of her masts, but the phrase seems to have originated in the north Atlantic, where, in summer, this would have typically been at about 11 a.m.. This was the time at which, by custom and rule, the first rum "tot" of the day was issued to officers and men (the officers had their tots neat, while the men had theirs diluted with water), hence its connection with taking one's first alcoholic drink of the day.

The earliest mention of this phrase collected by the OED is in Rudyard Kipling's *From Sea to Sea* in 1899, where it is used as a metaphor referring to drinking habits. However, the phrase was in use earlier, in the same context. One example is from the first volume of *Life,* from the issue of May 31, 1883.

## INVERTED BOW

In ship design, an inverted bow is a ship's or large boat's bow whose farthest forward point is not at the top. The result may somewhat resemble a submarine's bow. Inverted bow maximizes the length of waterline and hence the hull speed. Inverted bows were popular on battleships and large cruisers in the early 20th century. They fell out of favour as they were very wet on high speeds and heavy seas, but have made a comeback on modern ship design.

## WINGED KEEL

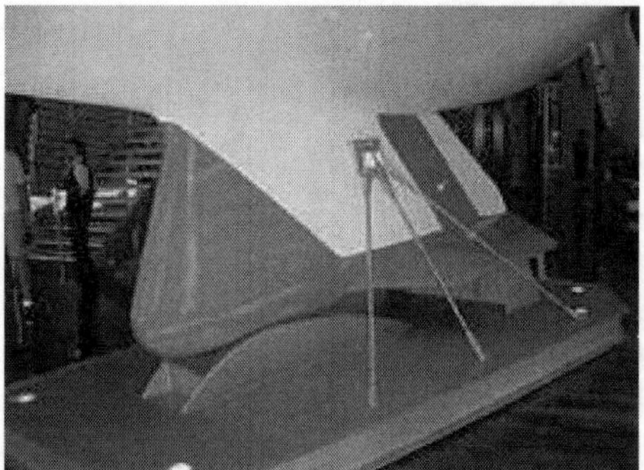

The Winged Keel is a sailboat keel originally designed by Ben Lexcen and made its first appearance on the 12-metre class yacht *Australia II* in the 1983 America's Cup. Along with Australia II's efficient sail design, this keel was one of the factors contributing to *Australia II's* success.

The increased stability afforded by the winged keel, due mainly to the extra lead in the wings producing a very low centre of gravity, allowed *Australia II* to be as short and light as possible under the 12 meter rules, and still carry enough sail to beat the American entry. The wings were angled downwards at about 20 degrees, which, since they were lifting downwards, acted as a dihedral. The lateral wings of a winged keel are usually of moderate aspect ratio, that uses a nearly horizontal foil, the "wing", at the bottom to provide additional span. Note that, contrary to classic configuration, the keel is "upside down" under the hull (the root chord is smaller than the tip (bottom) chord. Each wing acts as a winglet, effectively doubling the keel aspect ratio and reducing lift-induced drag.

Because the yacht is heeled over when sailing upwind the leeward foil is closer to vertical and provides additional side force, reducing the leeway angle, hence making the boat sail upwind more efficiently.

Winged keels are generally found on high performance sailboats, if they are not prohibited by class rules. They are only of benefit for yachts sailing upwind where stability and the ability to produce side force are important and if the draft is limited by the class rule. Downwind the extra skin friction drag is a hindrance. Besides the performance benefits, winged keels can also be applied to pleasure boats as a way to reduce draft, allowing for greater versatility when gunkholing.

Winged keels may incorporate a significant amount of ballast where the wings join the keel. This feature makes these winged keels also another type of bulb keel.

## WINDLASS ROOM

A windlass room is a triangular space enclosed within the bow of a freshwater ship where the anchor windlasses are located. Often, windlasses for handling docklines are located here as well. This room is where the forecastle of a saltwater ship would be located.

## WELL DOCK

In modern amphibious warfare usage, a well dock or well deck, officially termed a wet well in U.S. Navy instructions when the well deck is flooded for operations, is a hangar-like deck located at the waterline in the stern of some amphibious warfare ships. By taking on water the ship can lower its stern, flooding the well deck and allowing boats, amphibious vehicles and landing craft to dock within the ship.

### EVOLUTION OF STRUCTURE AND TERM

The structure on the modern naval amphibious ships does not exactly fit the traditional "well deck" definition of a weather, or open deck, that is lower than adjacent decks, surrounded by bulkheads that would, lacking proper drainage, form a catchment for water; however, the structure has its origins in such an exaggerated deep deck on World War II era tank landing craft (TLC) the British forces were considering.

On July 19, 1941 Major R. E. Holloway, Royal Engineers, forwarded a design from a 1924 patent by Otto Popper of the Danube International Commission concerning a barge transporter for Danube barges. That evolved into the British TLC-C that would become the Dock Landing Ship (LSD) that had an open, very deep and special purpose well deck open to the elements and thus technically a "well deck" in the traditional definition.

The open well deck began to see covering in part with platforms, particularly helicopter landing platforms, and in most modern form is no longer a weather deck, being completely enclosed overhead. The structure is seen in ships with a substantial air assault role such as the USS *Tarawa*.

The term "well deck" is still used for the newer structure in the U.S. Navy while other naval usage appears to be more commonly "well dock" that more clearly separates traditional usage from modern naval function.

### COMMERCIAL APPLICATION

Some commercial vessels have similar structures for purposes similar to the military versions. The German Baco Liner ships use bow doors. Previous barge carriers were of similar design to the military ships with a large stern well deck. The increased use of containers and container port facilities has decreased the use of this type of commercial vessel with their main use serving regions with less well developed ports.

## WELL DECK

In traditional nautical use well decks were decks lower than decks fore and aft, usually at the main deck level, so that breaks appear in the main deck profile as opposed to a flush deck profile. The term goes back to the days of sail. Late 20th Century commercial and military amphibious ships have applied the term to an entirely different type of hangar-like structure, evolving from exaggerated deep "well decks" of World War II amphibious vessels, that can be flooded for lighters or landing craft.

### TRADITIONAL

A well deck is an exposed deck (weather deck) lower than decks fore and aft. In particular it is one enclosed by bulwarks limiting flow of water and thus drainage so that design requirements are specific about drainage and maintenance of such drainage with that definition applying even to small vessels. The United States Coast Guard, Sector Upper Mississippi River, Small Passenger Vessel Information Package notes:

- ""Well deck" means a weather deck fitted with solid bulwarks that impede the drainage of water over the sides or an exposed recess in the weather deck extending 1/2 or more of the length of the vessel over the weather deck".

Explicit requirements exist for drainage requirements on ships with well decks. On many vessels the cargo hatches and cargo handling booms and winches were located on the well decks between a central superstructure and raised forecastle and poop.

Naval vessels into the World War II era often had well decks between stacks with the space sometimes reserved for aircraft and catapults as seen in the photograph of the USS *Indianapolis*'s well deck (note ladders down into the well deck from surrounding decks). In commercial ship design categories the "well deck ship" type the profile of the main deck line is broken rather than being flush or unbroken from bow to stern. The earliest cargo steamers were flush deck with openings below rimmed only with low casings. Later designs eliminated this design as dangerous with bulwarks and eventual evolution into the "three island" design with raised forecastle, central bridge and poop superstructures and well decks between.

The well decks of passenger liners were often for the lowest class passenger's use with the well deck of the *Titanic* reserved for third class passengers. By mid 20th Century the concept of well deck design in passenger liners was "old fashioned" and newer ships were designed with flush decks.

## MODERN MILITARY

In modern amphibious warfare usage, a well dock, is a hangar-like deck located at the waterline in the stern of some amphibious warfare ships. By taking on water the ship can lower its stern, flooding the well deck and allowing boats, amphibious vehicles and landing craft to dock within the ship. In the U.S. Navy, this is referred to as a well deck, and officially termed a wet well when the well deck is flooded for operations. The structure on the latest naval amphibious ships does not precisely fit the traditional meaning of a *weather*, or open deck, that is lower than adjacent decks, surrounded by bulkheads and lacking proper drainage would form a catchment for water; however the structure has its origins in such an exaggerated deep deck on World War II era tank landing craft (TLC) the British forces were considering.

On 19 July 1941 Major R. E. Holloway, Royal Engineers, forwarded a design from a 1924 patent by Otto Popper of the Danube International Commission concerning a barge transporter for Danube barges. The application of that idea to TLCs evolved into the British "Tank Landing Craft carrier" and that would become the Dock Landing Ship (LSD) that had an open, very deep and special purpose well deck open to the elements and thus technically a "well deck" in the traditional definition.

The structure underwent an evolutionary change to become an enclosed structure, essentially a floodable compartment, for docking amphibious vehicles that in most modern versions has lost its weather deck character even as the more modern commercial ship designs have abandoned the traditional well deck structures that are not typically features of today's bulk cargo ships, container ships or passenger ships.

### Commercial application

Some commercial vessels have similar structures for purposes similar to the military versions. The German Baco Liner ships use bow doors. Previous barge carriers were of similar design to the military ships with a large stern well deck. The increased use of containers and container port facilities has decreased the use of this type of commercial vessel with their main use serving regions with less well developed ports.

## WATERLINE LENGTH

The waterline length (originally Load Waterline Length, abbreviated to LWL) is a measurement of ships and boats. The term denotes the length of the vessel at the point where it sits in the water. It excludes the total length of the boat, such as features that are out of the water. Most boats rise outwards at the bow and stern, so a boat may be quite a bit longer than its waterline length. In a ship with such raked stems, naturally the waterline length changes as the draft of the ship changes, therefore it is measured from a defined loaded condition.

Length at the waterline is often abbreviated as lwl, w/l, w.l. or wl.

This measure is essential in determining a lot of properties of a vessel, such as how much water it displaces, where the bow and stern waves are, hull speed, amount of bottom-paint needed, etc. In sailing boats, as the wind increases heeling angle, the length on the waterline will typically increase. This will usually enable a greater maximum speed.

## WATER CHANNEL

A water channel is an experimental tank for studying resistance and propulsion behaviour of ships, submarines, or other sea vessels. In the study of naval architecture (ocean engineering), its general purpose is to allow understanding of the motion behavior of ships in advance.

## WALE

A wale is a thick plank of wood fastened to the side of a ship to provide protection from wear. The garboard is the wale next to the keel; the gunwale is the top such plank and covers the heads of the timbers between the main and fore drifts. A horizontal member of a tieback wall which transmits the force from the tieback to the beams.

## VENETIAN ARSENAL

The Venetian Arsenal (Italian: *Arsenale di Venezia*) was a complex of state-owned shipyards and armories clustered together in Venice in northern Italy. It was responsible for the bulk of Venice's naval power during the middle part of the second millennium AD. It was "one of the earliest large-scale industrial enterprises in history".Construction of the Arsenal began around 1104, during Venice's republican era. It became the largest industrial complex in Europe prior to the Industrial Revolution, spanning an area of about 45 ha

(110 acres), or about fifteen percent of Venice. Surrounded by a 2 mi (3.2 km) rampart, laborers and shipbuilders regularly worked within the Arsenal, building ships that sailed from the city's port. With high walls shielding the Arsenal from public view and guards protecting its perimeter, different areas of the Arsenal each produced a particular prefabricated ship part or other maritime implement, such as munitions, rope, and rigging. These parts could then be assembled into a ship in as little as one day. An exclusive forest owned by the Arsenal navy, in the Montello hills area of Veneto, provided the Arsenal's wood supply.

The Arsenal produced the majority of Venice's maritime trading vessels, which generated much of the city's economic wealth and power, lasting until the fall of the republic to Napoleon's conquest of the area in 1797. It is located in the Castello district of Venice, and it is now owned by the state.

The Byzantine-style establishment may have existed as early as the eighth century, though the present structure is usually said to have been begun in 1104 during the reign of Ordelafo Faliero, although there is no evidence for such a precise date. It definitely existed by the early 13th century.

Initially the state dockyard worked merely to maintain privately built naval ships, but in 1320 the *Arsenale Nuovo* was built, much larger than the original. It enabled all the state's navy and the larger merchant ships to be both constructed and maintained in one place. The Arsenal incidentally became an important center for rope manufacture, and housing for the arsenal workers grew up outside its walls.

Venice developed methods of mass-producing warships in the Arsenal, including the frame-first system to replace the Roman hull-first practice. This new system was much faster and required less wood. At the peak of its efficiency in the early sixteenth century, the Arsenal employed some 16,000 people who apparently were able to produce nearly one ship each day, and could fit out, arm, and provision a newly built galley with standardized parts on a production-line basis not seen again until the Industrial Revolution.

The staff of the Arsenal also developed new firearms at an early date, beginning with bombards in the 1370s and numerous small arms for use against the Genoese a few years later. The muzzle velocity of handguns was improved beyond that of the crossbow, creating armor-piercing rounds. Arsenal-produced arms were also noteworthy for their multi-purpose utility; the Venetian *condottieri* leader, Bartolomeo Colleoni, is usually given credit as being the first to mount the Arsenal's new lighter-weight artillery on mobile carriages for field use.

The Arsenal's main gate, the *Porta Magna,* was built around 1460 and was the first Classical revival structure built in Venice. It was perhaps built by Antonio Gambello from a design by Jacopo Bellini. Two lions taken from Greece situated beside it were added in 1687. One of the lions, known as the Piraeus Lion, has runic defacements carved in it by invading Scandinavian mercenaries during the 11th century. In the late 16th century, the Arsenal's designers experimented with larger ships as platforms for heavy naval guns. The largest was the galleass, already used at the Battle of Lepanto against the Ottoman Turks, and developed from the old merchanting "great galley". It was huge, propelled by both sails and oars, with guns mounted on wheeled carriages along the sides in the modern fashion.

It was slow and unwieldy in battle, however, and few were ever built. The galleon, also developed at the Arsenal, was an armed sailing ship, a slimmer version of the merchant "round ship". It was useful in major naval battles, but not in the small bays and off the extensive lee shores of the Dalmatian coast.

Significant parts of the Arsenal were destroyed under Napoleonic rule, and later rebuilt to enable the Arsenal's present use as a naval base. It is also used as a research center, an exhibition venue during the Venice Biennale and is home to a historic boat preservation center.

## GALILEO AND THE ARSENAL

In 1593, Galileo became an external consultant to the Arsenal, advising military engineers and instrument makers and helping to solve shipbuilders' problems, many of them relating to matters of ballistics. He was also responsible for creating some major innovations in the production and logistics of the Arsenal.

As a result of his interactions with the Arsenal, Galileo published a book later in his life addressing a new field of modern science, that concerned with the strength and resistance of materials. This science largely saw its roots in the knowledge of the shipwrights of the Venetian Arsenal. It is also supposed that Galileo's initial visits to the Arsenal were as a result of his initiative to further investigate Aristotle's questions concerning shipbuilding and navigation, found in the Mechanical Questions of Aristotle. As a result of these investigations, which were pursued by observing the work of the shipwrights, Galileo was asked to help in resolving a specific problem with the rowing units of the galleys.

As a result of his study of Aristotle, and in particular Question 4 regarding the propulsion of ships by oar, Galileo was able to produce a response to this question and ended up becoming a major source of information for the shipbuilders of the Arsenal concerning matters of rowing, instruments, and ballistics.

## VENICE'S NAVAL POWER

Venice's wealth and power rested in her ability to control trade in the Mediterranean. This would not have been possible without an extremely large navy and merchant force. By 1450, over 3,000 Venetian merchant ships were in operation, both as supply ships for Venetian merchants and as warships for the Venetian navy. This amazingly large amount of ships required constant maintenance and outfitting. The Venetian Arsenal was not only able to function as a major shipyard, but was also responsible for these routine maintenance stops that most Venetian galleys required. This required a large amount of money and the Venetian government spent almost 10% of its income on the Arsenal.

However, this naval power resulted in the domination of Mediterranean commerce. Venice's leading families, largely merchants and nobleman, were responsible for creating some of the grandest palaces and employing some of the most famous artists ever known. This opulence and wealth would not have been possible without the naval force constructed by the Arsenal. Indeed with the creation of the Great Galley and the mass production capacity of the Arsenal, "the fleets of Venice were the basis for the greatest commercial power the European world had yet seen."

## V-HULL (BOAT)

A V-hull, is the shape of a boat or ship in which the contours of the hull come in a straight line to the keel. V-hull designs are usually used in smaller boats and are useful in providing space for ballast inside the boat.

## TURRET SHIP

Turret ships were a 19th-century type of warship, the earliest to have their guns mounted in a revolving gun turret, instead of a broadside arrangement. The USS *Monitor*, built in 1862, is famous for, being the first ship mounted with a turret. Pressure from the Royal Navy's proponent of turrets, Captain Cowper Phipps Coles, led to the building of HMS *Captain* to his design in 1869. She was a low freeboard design and capsized during a heavy gale while under sail. HMS *Monarch* and HMVS *Cerberus* built at the same time proved more durable and HMS *Devastation* of 1871 led directly to the modern battleship. Mostly turret ships were monitors, oceangoing and coastal. Ships like the *Zhenyuan* were used in the Sino-Japanese War and Russo-Japanese War.

## TURRET DECK SHIP

A turret deck ship is a type of merchant ship with an unusual hull, designed and built in the late 19th and early 20th centuries.

The hulls of turret deck vessels were rounded and stepped inward above their waterlines. This gave some advantages in strength and allowed them to pay lower canal tolls under the tonnage measurement rules then in effect. The type ceased to be built after those rules changed.

### DEVELOPMENT

Turret deck ships were inspired by the visit of the U.S. whaleback vessel

SS *Charles W. Wetmore* to Liverpool in 1891. Like others of her type, *Wetmore* had a hull in the form of a flattened cigar, with a continuous curve above the waterline to where the sides met amidships. The superstructure atop the hull was in round or oval "turrets", so named because of their resemblance to gunhouses on contemporary warships.

In 1893 William Doxford and Sons Ltd. ("Doxford") of Sunderland, England built one whaleback under license from the type's designer, but had already built its first turret deck ship to a design by Arthur Havers, the concern's chief draughtsman.

Havers toned down the more radical features of the whaleback. His design retained conventional bows and sterns instead of the upswept conoid "snout" of the whaleback. Instead of a rounded hull, the hull of a turret vessel was stepped inward above the waterline, but the horizontal and vertical surfaces of the hull met in curves rather than by right angles as in conventional ships. Finally, the design joined the rounded turrets of whalebacks into one long and narrow rectangular structure (also called a "turret") of about half the beam, and used that space as part of the hold.

## DESCRIPTION AND DESIGN

In side profile, turret deck ships resembled other merchant vessels with flush decks or with small forecastles and poop decks. In cross-section the differences between turret deck vessels and more conventional ships are apparent. There was no gunwale; the vertical side of a turret ship curves inward above the load line to a horizontal plane. This flat area was known as the *harbour deck*. Further inboard, this "deck" arced to the vertical again by a reverse curve. That vertical plane then joined the weather deck atop the turret at a right angle. Structurally these elements were part of the hull, not of the superstructure, and the cargo holds of the ship extended up to the true weather deck atop the turret.

This design, and that of its near relative trunk deck ships, were said to maximize strength, allowing larger vessels and reduced the amount of steel needed for construction. In reality, it is more likely that the geometry inhibited the development of cracks in the sheer strake but vessels to this design were not any lighter than conventional vessels due to their unique geometry. In operation their hull form promoted self-trimming of homogenous cargo and inhibited shifting.

The design also called for a cellular double bottom, which was the probable reason for claims of the type's exceptional hull strength, but it also raised the centre of gravity of the cargo. A higher centre of gravity increased the roll period and reduced the violence of rolls. But loading heavy cargo too high, and failing to properly ballast the bottom tanks, raised the centre of gravity and led to instability. This led to accidents, a Board of Trade investigation, and cautions from Doxford on proper loading. The design also was inconvenient, as the narrowness of the turret made for smaller cargo hatches and restricted habitation spaces in the superstructure atop the turret.

Turret deck ships had a low net tonnage (an approximate measure of cargo space) in comparison to their deadweight capacity (weight of cargo) allowing them to operate at a lower fee structure than a conventional hull. Net tonnage is a computation of volume, and the method of measurement used at the Suez Canal to determine tolls was based on a measure of net tonnage which excluded some of the cargo spaces of these unconventional hulls. Turret and trunk deck ships therefore paid less in tolls than conventional ships of the same capacity.

In 1911 the toll measure changed at Suez to account for all cargo spaces, and contemporaneous refinements in the design of ships of more conventional hull form eliminated the structural advantages of turret deck ships. Construction of the type therefore ceased.

### HISTORY OF USE

Over 180 ships of the type had been built before the design was abandoned, 176 of them by William Doxford and Sons. They were used in both line voyage and tramp service until retired, wrecked, or lost in the First or Second World War. The Scotland-based Clan Line, which traded globally in cargos such as foodstuffs, timber, metals, manufactured goods, case oil, jute, tea, nitrates, and general cargo, used 30 of the type for those purposes.

While used for general freight, these ships were particularly suited to the carriage of bulk cargos such as grains, coal, and ores. Several were sold to Canadian interests for use in the latter trades on the Saint Lawrence River and Great Lakes of North America. The last of them, *Turret Cape*, operated until midcentury and was not scrapped until 1959.

## TUNNEL HULL

A tunnel hull is a type of boat hull that uses two typically planing hulls with a solid centre that traps air. This entrapment then creates aerodynamic lift in addition to the planing (hydrodynamic) lift from the hulls. Many times this is attributed to ground effect. Theoretical research and full-scale testing of tunnel hulls has demonstrated the dramatic contributions of 'close-proximity ground effect' on enhanced aerodynamic lift/drag in operation of performance tunnel hull designs.

Tunnel hulls are distinguishable from other catamarans by the typical close hull spacing and solid deck in between the hulls.

Formula 1 powerboats have a tunnel hull catamaran design allowing them to go faster. Tunnel hulls are a common design in offshore powerboat racing.

## TUMBLEHOME

In ship designing, the tumblehome is the narrowing of a ship's hull with greater distance above the water-line. Expressed more technically, it is present when the beam at the uppermost deck is less than the maximum beam of the vessel.

A small amount of tumblehome is normal in many designs in order to allow any small projections at deck level to clear wharves.

Tumblehome was common on wooden warships for centuries. In the era of oared combat ships it was quite common, placing the oar ports as far abeam as possible. This also made it more difficult to board by force, as the ships would come to contact at their widest points, with the decks some distance apart. The narrowing of the hull above this point made the boat more stable by lowering the weight above the waterline, which is one of the reasons it remained common during the age of cannon-armed ships.

In addition, the sloping sides of a tumblehome ship increased the effective thickness of the hull versus flat horizontal trajectory gunfire (a straight line through faced more material to penetrate) and increased the likelihood of a shell striking the hull being deflected—much the same reasons that later tank armor was sloped. It can be seen well in steel constructed warships of the early 1880s when the United States and most European navies began building steel warships.

France was predominantly strong in promoting the tumblehome design in their warships, advocating tumblehome reduced the weight of the upper deck, as well as making the vessels more seaworthy and creating greater freeboard. France sold their newly constructed pre-dreadnought battleship *Tsesarevich* to the Russian Imperial Navy in time for it to fight as Admiral Wilgelm Vitgeft's flagship at the Battle of the Yellow Sea on 10 August 1904. The Russo-Japanese War did prove however, that the controversial tumblehome battleships were excellent for transiting across the globe, especially when encountering narrow canals, and other waterways; but still could prove dangerously unstable when watertight integrity was breached.

However, the five follow-on tumblehome-designed *Borodino*-class battleships, which had been built in Russian yards to *Tsesarevich*'s basic design, fought the only decisive steel battleship fleet action in naval history on 27 May 1905 at Tsushima. The fact that three of the four (the fifth battleship, the *Slava* was not completed in time) "tumblehome" *Borodino*-class battleships were lost in this battle resulted in the discontinuing of the tumblehome design in future warships for nearly all navies.

A degree of tumblehome also facilitates paddling in a canoe or kayak, while a greater degree of flare (its opposite) accommodates more cargo.

## MODERN WARSHIP DESIGN

Tumblehome has been used in proposals for several modern United States Navy ship projects. The hullform also has an exaggerated ram bow, a wave-piercing bow shape where the stem rakes aft. This results in a more stable weapons platform, as the vessel does not rise over waves but passes through them. The rake of the stem is continued to the stern in the form of tumblehome. This combination of features results in a lower radar reflection.

This aspect gives the vessel something of the faceted appearance of a stealth aircraft. The ship is a compromise between a surface vessel and a modern-type submarine, which is designed to pass through the surface of the sea rather than over it.

However due to stability concerns most warships with wave-piercing hulls are multi-hull designs such as the HSV-2 Swift, USS Independence (LCS-2), Type 022 missile boat, etc.

The hull design is used on the DDG-1000 class (*Zumwalt*-class destroyer) of ships of the US Navy.

## IN NARROWBOAT DESIGN

The inward slope of a narrowboat's superstructure (from gunwales to roof) is referred to as tumblehome. The amount of tumblehome is one of the key design choices when specifying a narrowboat, because the widest part of a narrowboat is rarely more than 7 feet across, so even a modest change to the slope of the cabin sides makes a significant difference to the "full-height" width of the cabin interior. Too great a tumblehome would make a boat difficult to pass through for a tall person; too little and the cabin roof edges are at risk of damage when the boat is passing through a tunnel (many canal tunnels on the British inland waterways have subsided, bringing the curve of the roof closer to the water level).

## IN AUTOMOBILE DESIGN

The inward slope of the "greenhouse" above the beltline is also called the tumblehome. Less commonly, the inward curve of the body near the bottom may also be called a tumblehome. In 21st century automobile designs this *turnunder* is less pronounced or eliminated to reduce aerodynamic drag and to help keep the lower portions of the vehicle cleaner under wet conditions.

## IN RAILWAY DESIGN

The inwardly curving portions of railway passenger carriages at the point where the carriage sides join the underframes is also called the tumblehome. Tumblehome styling of railway carriages was particularly prevalent in the British Isles (or on railways influenced by British engineers or equipment builders) in the 19th century and "wood body" era of the early 20th century. This enabled the wooden step running the length of the carriage to remain within the dimensions of the loading gauge, while allowing maximum width for the main body of the carriage. Thus there was space to place a foot when entering or leaving the carriage.

## TRIBON

Tribon is a naval architecture program originally developed by Kockum Computer Systems (KCS) for designing commercial and naval vessels. KCS was spun off from Kockums shipyards as an independent company, later renamed Tribon Systems, which was in turn acquired by AVEVA in 2004.

Tribon is actually a family of programs that create and refer to a common set of databases containing the design details of the ship. Together, these databases are used to depict a 3D model of the ship, with embedded information for all of the parts of the design, from ship structural elements to pipe segments to equipment. Many common ship design elements, especially equipment, can be downloaded from databases on Tribon.com.

## TREENAIL

A treenail, also trenail or trunnel, is a wooden peg or dowel used to fasten pieces of wood together, especially in timber frame construction and wooden shipbuilding. It is an ancient technology. Covered bridges in the U.S. often use treenails as fasteners. Many such bridges are still in use. Locust is a favorite wood when making trunnels in shipbuilding due to its strength and rot resistance.

A method of firmly securing such a fastener was to cut a parallel peg of a softer wood, and then expand its outer end with a wedge of much harder wood driven into it.

Ancient shipbuilding used treenails to bind the boat together. They had the advantage of not giving rise to "nail-sickness", a term for decay accelerated and concentrated around metal fasteners. Increased water content causes wood to expand, so that treenails gripped the planks tighter as they absorbed water. Similar wooden trenail fastenings were used as alternatives to metal spikes to secure rail-support "chairs" to wooden sleepers in early Victorian times.

## TRAILBOARD

The trailboards are a pair of boards that may be found at the bow of certain sailing vessels, where they run from the figurehead or billethead back to or towards the hawsepipe. They are in the main decorative, though they often bear the name of the ship; they may be more or less elaborately carved and painted.

## SIGNIFICANT WAVE HEIGHT

In physical oceanography, the significant wave height (SWH or $H_s$) is defined traditionally as the mean wave height (trough to crest) of the highest third of the waves ($H_{1/3}$). Nowadays it is usually defined as four times the standard deviation of the surface elevation – or equivalently as four times the square root of the zeroth-order moment (area) of the wave spectrum. The symbol $H_{m0}$ is usually used for that latter definition. The significant wave height may thus refer to $H_{m0}$ or $H_{1/3}$; the difference in magnitude between the two definitions is only a few percent.

The original definition resulted from work by the oceanographer Walter Munk during World War II. The significant wave height was intended to mathematically express the height estimated by a "trained observer". It is commonly used as a measure of the height of ocean waves.

## STATISTICAL DISTRIBUTION OF THE HEIGHTS OF INDIVIDUAL WAVES

Significant wave height, scientifically represented as $H_s$ or $H_{sig}$, is an important parameter for the statistical distribution of ocean waves. The most common waves are less in height than $H_s$. This implies that encountering the significant wave is not too frequent. However, statistically, it is possible to encounter a wave that is much higher than the significant wave.

Generally, the statistical distribution of the individual wave heights is well approximated by a Rayleigh distribution. For example, given that $H_s$ is 10 metres (33 feet), statistically:

- 1 in 10 will be larger than 10.7 metres (35 ft)
- 1 in 100 will be larger than 15.1 metres (50 ft)
- 1 in 1000 will be larger than 18.6 metres (61 ft)

This implies that one might encounter a wave that is roughly double the significant wave height. However, in rapidly changing conditions, the disparity between the significant wave height and the largest individual waves might be even larger.

### Other statistics

Other statistical measures of the wave height are also widely used. The RMS wave height, which is defined as square root of the average of the squares of all wave heights, is approximately equal to $H_s$ divided by 1.4.

For example, according to the Irish Marine Institute:

- "... at midnight on 9/12/2007 a record significant wave height was recorded of 17.2m at with a period of 14 seconds."

## MEASUREMENT

Although most measuring devices estimate the significant wave height from a wave spectrum, satellite radar altimeters are unique in measuring directly the significant wave height thanks to the different time of return from wave crests and troughs within the area illuminated by the radar. The maximum ever measured wave height from a satellite is 20.1m during a North Atlantic storm in 2011.

## WEATHER FORECASTS

The World Meteorological Organization stipulates that certain countries are responsible for providing weather forecasts for the world's oceans. These respective countries' meteorological offices are called Regional Specialized Meteorological Centers, or RSMCs. In their weather products, they give ocean wave height forecasts in significant wave height. In the United States, NOAA's National Weather Service is the RSMC for a portion of the North Atlantic, and a portion of the North Pacific.

The Ocean Prediction Center and the Tropical Prediction Center's Tropical Analysis and Forecast Branch (TAFB) issue these forecasts. RSMCs use wind-wave models as tools to help predict the sea conditions. In the U.S., NOAA's WAVEWATCH III model is used heavily.

### GENERALIZATION TO WAVE SYSTEMS

A *significant wave height* is also defined similarly, from the *wave spectrum*, for the different systems that make up the sea. We then have a *significant wave height* for the wind-sea or for a particular swell.

## SHIPYARD TRANSPORTER

A Shipyard transporter is a heavy equipment transporter or heavy trailer and is often used in shipyards but is not limited to them. As its name implies, a shipyard transporter is often used in transporting ship sections from the shipbuilder workshop to the dock to assemble a whole vessel. Shipyard transporters adopt hydraulic suspension and steering to achieve loading, moving and steering.

As each wheel has an independent steering unit, the shipyard transporter can achieve 8 direction moving and turning 360 degrees around the middle point of the transporter as the origin point. Some times, Self Propelled Modular Transporters (SPMT) or self-propelled trailers are also called shipyard transporters.

## SHIPYARD

Shipyards and dockyards are places where ships are repaired and built. These can be yachts, military vessels, cruise liners or other cargo or passenger ships. Dockyards are sometimes more associated with maintenance and basing activities than shipyards, which are sometimes associated more with initial construction. The terms are routinely used interchangeably, in part because the evolution of dockyards and shipyards has often caused them to change or merge roles.

Countries with large shipbuilding industries include South Korea, Australia, Japan, China, Germany, Turkey, Poland and Croatia. The shipbuilding industry tends to be more fragmented in Europe than in Asia. In European countries there are more smaller companies, compared to the fewer, larger companies in the shipbuilding countries of Asia. Most shipbuilders in the United States are privately owned, the largest being

Huntington Ingalls Industries, a multi-billion dollar defense contractor. The publicly owned shipyards in the US are Naval facilities providing basing, support and repair.

Shipyards are constructed by the sea or by tidal rivers to allow easy access for their ships. In the United Kingdom, for example, shipyards were established on the River Thames (King Henry VIII founded yards at Woolwich and Deptford in 1512 and 1513 respectively), River Mersey, River Tees, River Tyne, River Wear and River Clyde – the latter growing to be the World's pre-eminent shipbuilding centre.

Sir Alfred Yarrow established his yard by the Thames in London's Docklands in the late 19th century before moving it northwards to the banks of the Clyde at Scotstoun (1906–08). Other famous UK shipyards include the Harland and Wolff yard in Belfast, Northern Ireland, where the *Titanic* was built, and the naval dockyard at Chatham, England on the Medway in north Kent.

The site of a large shipyard will contain many specialised cranes, dry docks, slipways, dust-free warehouses, painting facilities and extremely large areas for fabrication of the ships.

After a ship's useful life is over, it makes its final voyage to a shipbreaking yard, often on a beach in South Asia. Historically shipbreaking was carried on in drydock in developed countries, but high wages and environmental regulations have resulted in movement of the industry to developing regions.

The world's earliest known dockyards were built in the Harappan port city of Lothal circa 2400 BC in Gujarat, India. Lothal's dockyards connected to an ancient course of the Sabarmati river on the trade route between Harappan cities in Sindh and the peninsula of Saurashtra when the surrounding Kutch desert was a part of the Arabian Sea.

Lothal engineers accorded high priority to the creation of a dockyard and a warehouse to serve the purposes of naval trade. The dock was built on the eastern flank of the town, and is regarded by archaeologists as an engineering feat of the highest order. It was located away from the main current of the river to avoid silting, but provided access to ships in high tide as well.

The name of the ancient Greek city of Naupactus means "shipyard" (combination of the Greek words íáýò *naus* ship, boat and ðÞãíõìé *pêgnumi, pegnymi* builder, fixer). Naupactus' reputation in this field extends to the time of legend, where it is depicted as the place where the Heraclidae built a fleet to invade the Peloponnesus.

In the Spanish city of Barcelona, the Drassanes shipyards were active from at least the mid-13th century until the 18th century, although it at times served as a barracks for troops as well as an arsenal. During its time of operation it was continuously changed, rebuilt and modified, but two original towers and part of the original eight construction naves remain today. It is currently a maritime museum.

Ships were the first items to be manufactured in a factory, several hundred years before the Industrial Revolution, in the Venice Arsenal, Venice, Italy. The Arsenal apparently mass-produced nearly one ship every day using pre-manufactured parts, and assembly lines and, at its height, employed 16,000 people.

## WORSHIPFUL COMPANY OF SHIPWRIGHTS

The Worshipful Company of Shipwrights is one of the livery companies of the City of London.

The Shipwrights' Company, unlike other livery companies, has not received a royal charter because maritime trade by definition was never confined within the boundaries of the Square Mile; instead a corporate body of London shipwrights grew over time, their first recorded reference being in the twelfth century; thus the company's status is considered as being incorporated "by prescription".

By contrast a royal charter *was* issued in 1612 to the "Master, Wardens, and Commonalty of the Art or Mystery of Shipwrights of Redriff in the County of Surrey": *i.e.* Rotherhithe. This led to a dispute about jurisdiction between the two companies, which was resolved in 1684 when the *Rotherhithe "charter"* was cancelled.

The Shipwrights' Company received confirmation of its City of London livery status in 1782. Ever since Queen Victoria's reign the company continues to enjoy a special connection with the Royal Family, several of whom are liverymen today; Prince Charles was installed on 10 May 2011 as prime warden and served for 2011-12 before succeeding his father, the Duke of Edinburgh, as permanent master shipwright on 16 February 2012.

Although the Shipwrights' Company is no longer a trade association solely representing the shipbuilding industry, its links with maritime, shipping and global trade remain strong. Whilst sons and daughters of members can join as freemen of the company, only those with a professional maritime background (military or commercial) may become liverymen.

The company supports maritime research, numerous charities, as well as the work of the Lord Mayor of London, the City of London Corporation and the Sheriffs of the City of London.

The company is also involved in charitable educational activities, which includes the support of the London Nautical School and George Green's School in London.

The current clerk to the Shipwrights' Company is Lt Col Andrew Milne RM, whose role combines that of executive officer as well as supporting the Prime Warden: Simon Robinson. Its honorary chaplain is the Revd Canon Bill Christianson. The company ranks fifty-ninth in the City livery order of precedence and cohabits with the Ironmongers' Company. Its motto is "Within The Ark Safe For Ever".

## SHIPLIFT

A shiplift is a modern alternative for a slipway, a floating dry dock or a graving dry dock. A shiplift is used to dry dock and launch ships. It consists of a structural platform that is lifted and lowered exactly vertical, synchronously by a number of hoists. First, the platform is lowered underwater, then the ship is floated above the support, and finally the platform with shipsupport and ship is lifted and the ship is brought to the level of the quay.

Nowadays, shiplift are most of the time supplied under rules of a classification authority. Lloyd's Register of Shipping is the authority with most experience in the certification or classification of shiplifts. Whereas "shiplift" is the word that is normally used, the term used by Lloyd's register is "Mechanical Lift Dock".

It is important to synchronize the winches. Modern systems use electrical control and drive systems for the winches. Old fashioned electrically driven shiplift systems needed so much starting current that they often could only be used in the weekend when other electrical equipment was not in use. The modern shiplifts however do not have this disadvantage; they can start the movement in any desired curve.

Many shiplifts use a transfer system for ships so that the vessels can be transported from the water to a parking place where they can be painted or repaired. One shiplift can serve many parking places, while a dry docking installation can only dock one ship.

For large vessels the transfer system consists of a number of trolleys or cradles, supported by high capacity steel wheels. The wheels drive on heavy duty rails.

The transport can be one directional, but in order to serve more parking places, two directional systems are used. These two directional systems make use of a *traverser carriage* or sometimes the wheels of the trolleys can be turned over 90 degrees. Above the wheels often hydraulic cylinders are located so that the ship can be transported upon a so-called "fluid bed". The cylinders are also used to lower the ships on a static support, so that the transfer system with the active elements can be recovered.

### PLEASURE CRAFT

Smaller devices called "Boat lifts" are available that simply raise yachts, sailboats, or small watercraft above the water level at a dock for storage, for reduced maintenance cost and increased security. They can be operated by cables, by hydraulics, or by air flotation and some are "drive on" to segmented flexing supports.

Even smaller, are devices sometimes called "Portable boat lifts" which are easily transported cranes that allow one person to separate a boat from a trailer by themselves.

## SHIPFITTER

A shipfitter is a marine occupational classification used both by naval activities and among ship builders; however, the term applies mostly to certain workers at commercial and naval shipyards during the construction or repair phase of a ship. The term is derived from the words "ship" and "fit" — essentially, "fitting" parts of the "ship" together.

### NAVAL USAGE

A shipfitter is an enlisted or civilian person who works on heavy metal like high-tensile steel, high yield strength steel, etc. Shipfitters fabricate, assemble and erect all structural parts of a ship, coordinate all fixed tank work performed on submarines and ships, and coordinate all sonar dome work. Shipfitters also use heavy machinery such as plate planners, shears, punches, drill presses, bending rolls, bending slabs, plate beveler, saws, presses up to 750 tons, angle roll (vertical and horizontal), dogs and wedges. Shipfitters are responsible for hydro and air testing of tanks and compartments as well as perform grinding, drilling, and fit up operations on submarines and surface crafts.

### SHIP BUILDER USAGE

A shipfitter is a worker who "fits" together the structural portions of a ship by either welding or by riveting.

## SHIPBUILDING CONTRACT

Shipbuilding contract, which is the contract for the complete construction of a ship, concerns the sales of future goods, so the property could not pass title at the time when the contract is concluded. The aim of shipbuilding contract is to regulate a substantial and complex project which the builders and buyers assume long-term obligations to other and bear significant commercial risks.

Shipbuilding contract is a non-maritime contract and not within the Admiralty jurisdiction because it is insufficiently related to any rights and duties pertaining to sea commerce and/or navigation. The property passes to the buyer when the ship has been completed. To avoid difficulties, provision can be made for the property to pass in stage in the process of development and construction. It is different from most hire-purchase agreements where the seller has ownership of the property until the payment of the final installment.

Under the Sale of Goods Act 1979, this kind of agreement to sell 'future' goods may be a sale either by description or by sample. The sale of new building ship, which is large manufacturing project, is obviously undertaken by description. It is a condition to comply with the agreed description when performing the contract.

## STANDARD FORMS OF CONTRACT

Shipbuilding contract are constructed within the framework of standard contract forms amended by the contractual parties to meet their particular requirements. The choice of form will be based on the influence of trade association which the builders belong to.

Principal Form

- SAJ Form
  It is published by the Shipbuilders' Association of Japan in January 1974 and the framework of this form is commonly used in South Korea, China, Singapore and Taiwan.
- AWES Form
  It is the standard shipbuilding contract of the Association of European Shipbuilders and Shiprepairers which revised and reissued in May 1999.

National Form

- The Norwegian Shipowners' Association and Norwegian Shipbuilders' Association
- MARAD Form (The Maritime Administration of the United States Department of Commerce)
  It is used in relation to American newbuildings financed under Federal Ship Financing Program authorized by Title XI of the Merchant Marine Act 1936.

## FORMATION OF CONTRACT

There is no requirement that a shipbuilding contract should be concluded in writing, it will be also legally enforceable in oral provided that necessary formal elements are present. The main terms of an agreement, such as expenditure, timescale and risks involved in shipbuilding, are better to record in written form.

In order to create an enforceable agreement, the essential elements for an agreement to form a legally binding contract must be presented, they are:

- offer;
- acceptance;
- consideration;
- privity of contract
- intention to form a contract;
- capacity;

Where all these elements are present, a legally binding contract comes into effect. Otherwise, if any of the elements are missing, there is no legally binding contract.

## SLAMMING

Slamming is the impact of the bottom structure of a ship onto the sea surface. It is mainly observed while sailing in waves, when the bow raises from the water and subsequently impacts on it. Slamming induces extremely high loads to ship structures and is taken into consideration when designing ships.

## SLIPWAY

A slipway, also known as boat ramp or launch, is a ramp on the shore by which ships or boats can be moved to and from the water. They are used for building and repairing ships and boats. They are also used for launching and retrieving small boats on trailers towed by automobiles and flying boats on their undercarriage.

The nautical term ways is an alternative name for slipway. A ship undergoing construction in a shipyard is said to be *on the ways*. If a ship is scrapped there, she is said to be *broken up in the ways*.

As the word "slip" implies, the ships or boats are moved over the ramp, by way of crane or fork lift, prior to the move the vessel's hull is coated with grease, which then allows the ship or boat to "slip" off of the ramp and progress safely into the water.

Slipways are used to launch (newly built) large ships, but can only dry-dock or repair smaller ships. Pulling large ships against the greased ramp would require too much force. For dry-docking large ships, one must use carriages supported by wheels or by roller-pallets. These types of dry-docking installations are called "marine railways". Nevertheless the words "slip" and "slipway" are also used for all dry-docking installations that use a ramp.

### SIMPLE SLIPWAYS

In its simplest form, a slipway is a plain ramp, typically made of concrete, steel, stone or even wood. The height of the tide can limit the usability of a slip: unless the ramp continues well below the low water level it may not be usable at low tide. Normally there is a flat paved area on the landward end.

When used for building and repairing boats or small ships (i.e. ships of no more than about 300 tons), the vessel is moved on a wheeled carriage, which is run down the ramp until the vessel can float on or off the carriage. Such slipways are used for repair as well as for putting newly built vessels in the water.

When used for launching and retrieving small boats, the trailer is placed in the water. The boat may be either floated on and off the trailer or pulled off. When recovering the boat from the water, it is winched back up the trailer.

### WHALING

Whaling ships are usually equipped with a slipway at the back, to assist in hauling harpooned whales onto the main deck, where they are usually flensed.

## LIFEBOAT SLIPWAYS

To achieve a safe launch of some types of land-based lifeboats in bad weather and difficult sea conditions, the lifeboat and slipway are designed so that the lifeboat slides down a relatively steep steel slip under gravity. It is winched back up afterward. The Royal National Lifeboat Institution in the United Kingdom currently operates three different classes of lifeboat from its slipways: the Tyne, Mersey and, most recently, the Tamar.

## SHIP CONSTRUCTION

For large ships, slipways are only used in construction of the vessel. Normally they are arranged perpendicular to the shore line (or as nearly so as the water and maximum length of vessel allows) and the ship is built with its stern facing the water. Modern slipways take the form of a reinforced concrete mat of sufficient strength to support the vessel, with two "barricades" that extend to well below the water level taking into account tidal variations. The barricades support the two launch ways.

The vessel is built upon temporary cribbing that is arranged to give access to the hull's outer bottom, and to allow the launchways to be erected under the complete hull. When it is time to prepare for launching a pair of standing ways are erected under the hull and out onto the barricades. The surface of these ways are greased. (Tallow and whale oil were used as grease in sailing ship days.) A pair of sliding ways is placed on top, under the hull, and a launch cradle with bow and stern poppets is erected on these sliding ways.

The weight of the hull is then transferred from the build cribbing onto the launch cradle. Provision is made to hold the vessel in place and then release it at the appropriate moment in the launching ceremony, these are either a weak link designed to be cut at a signal or a mechanical trigger controlled by a switch from the ceremonial platform.

The process of transferring the vessel to the water is known as *launching* and is normally a ceremonial and celebratory occasion. It is the point where the vessel is formally named. At this point the hull is complete and the propellers and associated shafting are in place, but dependent on the depth of water, stability and weight the engines might have not been fitted or the superstructure may not be completed.

On launching, the vessel slides backwards down the slipway on the ways until it floats by itself.

Some slipways are built so that the vessel is side on to the water and is launched sideways. This is done where the limitations of the water channel would not allow lengthwise launching, but occupies a much greater length of shore. The *Great Eastern* built by Brunel was built this way as were many landing craft during World War II. This method requires many more sets of ways to support the weight of the ship. In both cases heavy chains are attached to the ship and the drag effect is used to slow the vessel once afloat until tugboats can move the hull to a jetty for fitting out. The practice of building on a slipway is dying out with the very large vessels introduced from about 1970.

Part of the reason is the space requirement for slowing and maneuvering the vessel immediately after it has left the slipway, but the sheer size of the vessel causes design problems, since the hull is basically supported only at its end points during the launch process and this imposes stresses not met during normal operation.

## SPONSON

Sponsons are projections from the sides of a watercraft, for protection, stability, or the mounting of equipment such as armaments or lifeboats, etc. They extend a hull dimension at or below the waterline and serve to increase flotation or add lift when underway.

Sponsons are commonly used on jetskis and other personal watercraft such as canoes to provide either additional buoyancy and thus stability against capsizing, or hydrodynamic forces to resist capsizing. They can often be easily attached to an existing craft in order to improve its stability.

They are far less common on ships than such stabilizing means as pontoons, outriggers, and dual hulls due to their comparatively poor performance in stabilizing large hulls. Sponsons are sometimes added to improve stability when ships are modified.

Sponsons are used on the fuselages of flying boats, as pioneered by German aerospace engineer Claudius Dornier during World War I. They take the form of a short wing which when travelling through the water provides hydrodynamic stability during take off and landing.

They are often used in larger helicopters where the internal space of the sponson can be used for fuel or to house landing gear without reducing cargo or passenger space in the fuselage as, for example, with the Sikorsky S-92 and the Bell 222.

By extension, the term "sponson" is sometimes used to describe similar projections from the sides of land vehicles, such as the World War I Mark I tank. In the case of the Bradley Fighting Vehicle, which doubles as a troop transport and armament platform, the sponson refers to the aspect of the vehicle body directly over the tracks and includes layers of hardened, bullet-proof materials to protect the occupants. It is also used as a storage space for both vehicle equipment and components, and either ammunition or crew belongings.

## STEERAGE (SHIP)

Steerage is the act of steering a ship. The rudder of a vessel can steer the ship only when water is passing over it. Hence, when a ship is not moving relative to the water it is in or cannot move its rudder, it does not respond to the helm and is said to have "lost steerage." The motion of a ship through the water is known as "making way." When a vessel is moving fast enough through the water that it turns in response to the helm, it is said to have "steerage way." That is why boats on rivers must always be under propulsion, even when traveling downstream.

## STEERING POLE

A steering pole is a light spar extending from the bow of a straight deck ship which aids the wheelsman in steering. Ancient literature indicates that steering poles have long been part of boat construction, and are referred to in ancient texts such as the Epic of Gilgamesh.

## STEM (SHIP)

The stem is the very most forward part of a boat or ship's bow and is an extension of the keel itself. The stem is the curved edge stretching from the keel below, up to the gunwale of the boat. The stem is more often found on wooden boats or ships, but not exclusively. The stem is part of the physical structure of a wooden boat or ship that gives it strength at the critical section of the structure, bringing together the port and starboard side planks of the hull.

### STYLES

There are two styles of stems: plumb and raked. When the stem comes up from the water, if it is perpendicular to the waterline it is "plumb." If it is inclined at an angle to the waterline it is "raked." (E.g., "The hull is single decked and characterized by a plumb stem, full bows, straight keel, moderate deadrise, and an easy turn of bilge.")

### STEMHEAD

Because the stem is very sturdy, the top end of it may have something attached, either ornamental or functional in nature. On smaller vessels, this might be a simple wood carving (ornamental) or cleat (functional). On large wooden ships, figureheads can be attached to the upper end of the stem.

## STERN

The stern is the rear or aft-most part of a ship or boat, technically defined as the area built up over the sternpost, extending upwards from the counter rail to the taffrail. The stern lies opposite of the bow, the foremost part of a ship. Originally, the term only referred to the aft port section of the ship, but eventually came to refer to the entire back of a vessel. The stern end of a ship is indicated with a white navigation light at night.

Sterns on European and American wooden sailing ships began with two principal forms: the *square* or *transom* stern and the *elliptical, fantail,* or *merchant* stern, and were developed in that order. The hull sections of a sailing ship located before the stern are composed of a series of U-shaped rib-like frames set in a sloped or "cant" arrangement, with the last frame before the stern being called the *fashion timber(s)* or *fashion piece(s)*, so called for "fashioning" the after part of the ship. This frame is designed to support the various beams that make up the stern.

In 1817 the British naval architect Sir Robert Seppings first introduced the concept of the round or *circular* stern. The square stern had been an easy target for enemy cannon, and could not support the weight of heavy stern chase guns. But Seppings' design left the rudder head exposed, and was regarded by many as simply ugly—no American warships were designed with such sterns, and the round stern was quickly superseded by the elliptical stern. The United States began building the first elliptical stern warship in 1820, a decade before the British.

The USS *Brandywine* became the first sailing ship to sport such a stern. Though a great improvement over the transom stern in terms of its vulnerability to attack when under fire, elliptical sterns still had obvious weaknesses which the next major stern development—the iron-hulled cruiser stern—addressed far better and with much different materials.

## TRANSOM STERN

In naval architecture, the term "transom" has two meanings. First, a transom can refer to any of the individual beams that run side-to-side or "athwart" the hull at any point abaft the fashion timber; second, a transom can refer specifically to the flat or slightly curved surface that is the very back panel of a transom stern. In this sense, a transom stern is the product of the use of a series of transoms, and hence the two terms have blended.

The stern of a classical sailing ship housed the captain's quarters and became increasingly large and elaborate between the 15th and 18th centuries, especially in the baroque era, when such wedding-cake-like structures became so heavy that crews sometimes threw the decoration overboard rather than be burdened with its useless weight. But until a new form of stern appeared in the 19th century, the transom stern was a floating house—and required just as many timbers, walls, windows, and frames. The stern frame provided the foundational structure of the transom stern, and was composed of the sternpost, wing transom, and fashion piece.

Abaft the fashion timber, the transom stern was composed of two different kinds of timbers:

- Transoms – These timbers extend across the low parts of the hull near the rudder, and are secured (either notched and/or bolted) to the sternpost. The transom located at the base of the stern, and the uppermost of the main transoms, was typically called the *wing* transom; the principal transom below this and level with the lower deck was called the *deck* transom; between these two were a series of *filling* transoms. If the stern had transoms above the wing transom, they would no longer be affixed to the sternpost. The first of these might be called a *counter* transom; next up was the *window sill* transom; above that, the *spar deck* transom. The larger the vessel, the more numerous and wider the transoms required to support its stern.

- Stern timbers (also called *stern frames*) – These timbers are mounted vertically in a series; each timber typically rests or "steps" on the wing transom and then stretches out (aft) and upward. Those not reaching all the way to the taffrail are called *short stern timbers,* while those that do are called *long stern timbers.* The two outermost of these timbers, located at the corners of the stern, are called the *side-counter timbers* or *outer stern timbers.* It is the stern timbers collectively which determine the backward slope of the square stern, called its rake.

The flat surface of any transom stern may begin either at or above the waterline of the vessel. The geometric line which stretches from the wing transom to the archboard is called the *counter*; a large vessel may have two such counters, called a *lower* counter and a *second* or *upper* counter. The lower counter stretches from directly above the wing transom to the lower counter rail, and the upper counter from the lower counter rail to the upper counter rail, immediately under the stern's lowest set of windows (called "lights").

## ELLIPTICAL STERN

The visual unpopularity of Seppings' circular stern was soon rectified by Sir William Symonds. In this revised stern, a set of straight post timbers (also called "whiskers", "horn timbers", or "fan tail timbers") stretches from the keel diagonally aft and upward.

It rests on the top of the sternpost and runs on either side of the rudder post (thus creating the "helm port" through which the rudder passes) to a point well above the vessel's waterline. Whereas the timbers of the transom stern all heeled on the wing transom, the timbers of the elliptical stern all heel on the whiskers, to which they are affixed at a 45° angle (i.e., "canted") when viewed from overhead and decrease in length as they are installed aft until the curvature is complete.

The finished stern has a continuous curved edge around the outside and is raked aft. Other names for the elliptical stern include a "counter stern," in reference to its very long counter, and a "cutaway stern." The elliptical stern began use during the age of sail, but remained very popular for both merchant and warships well into the nautical age of steam and through the first eight decades of steamship construction (roughly 1840–1920), despite the fact that the design left the rudder exposed and vulnerable in combat situations.

## CRUISER STERN

As ships of wooden construction gave way to iron and steel, the cruiser stern—another design without transoms and known variously as the *canoe* stern, *parabolic* stern, and the *double-ended* stern—became the next prominent development in ship stern design, particularly in warships of the latter half of the 20th century. The intent of this re-design was to protect the steering gear by bringing it below the armor deck.

The stern now came to a point rather than a flat panel or a gentle curve, and the counter reached from the sternpost all the way to the taffrail in a continuous arch. It was soon discovered that vessels with cruiser sterns experienced less water resistance when under way than those with elliptical sterns, and between World War I and World War II most merchant ship designs soon followed suit.

None of these three main types of stern has vanished from the modern naval architectural repertoire, and all three continue to be utilized in one form or another by different sets of designers and for a broad spectrum of uses. Variations on these basic designs have resulted in an outflow of "new" stern types and names, only some of which are itemized here.

## OTHER TYPES OF STERN

The reverse stern, reverse transom stern, sugar-scoop, or *retroussé* stern is a kind of transom stern that is raked backwards (common on modern yachts, rare on vessels before the 20th century); the vertical transom stern or plumb stern is raked neither forward nor back, but falls directly from the taffrail down to the wing transom. The rocket ship stern is a term for an extremely angled *retroussé* stern.

A double ended ship with a very narrow square counter formed from the bulwarks or upper deck above the head of the rudder is said to have a pink stern or pinky stern. The torpedo stern or torpedo-boat stern describes a kind of stern with a low rounded shape that is nearly flat at the waterline, but which then slopes upward in a conical fashion towards the deck (practical for small high-speed power boats with very shallow drafts).

A Constanzi stern is a type of stern designed for use on ocean-going vessels. It is a compromise between the 'spoon-shaped' stern usually found on ocean liners, and the flat transom, often required for fitting azimuth thrusters. The design allows for improved seagoing characteristics. It is the stern design on *Queen Mary 2*, and was originally proposed for SS *Oceanic* and *Eugenio C*, both constructed in the 1960s.

A lute stern is to be found on inshore craft on the Sussex, England shore. It comprises a watertight transom with the topside planking extended aft to form a non-watertight counter which is boarded across the fashion timbers curving outward aft from the transom. Some working boats and modern replicas have a similar form of counter, built to be water tight as described in the "transom stern" section above. These are being confused with lute sterns but as a lute is not watertight, a better term is needed. Chappelle in American Small Sailing Craft refers to a Bermudan boat with this form of counter, using the term "square tuck stern" to describe it. The term "tuck" is used in the northwest of England for this area of the hull at the sternpost, and for the bulkhead across the counter if one is fitted.

A bustle stern refers to any kind of stern (transom, elliptical, etc.) that has a large "bustle" or blister at the waterline below the stern to prevent the stern from "squatting" when getting underway. It only appears in sailboats, never in power-driven craft.

## STERNPOST

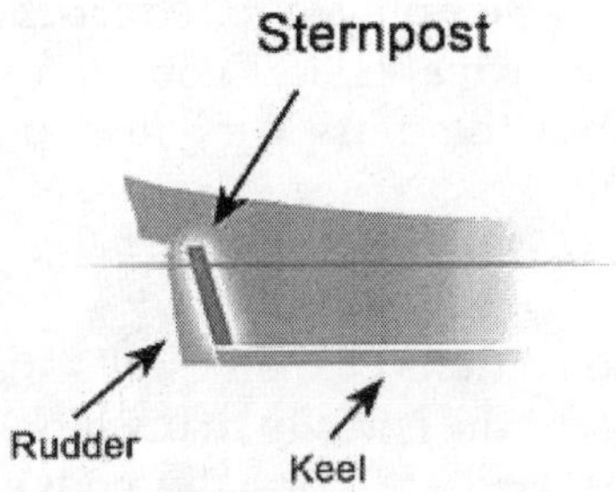


0

A sternpost is the upright structural member or post at the stern of a (generally wooden) ship or a boat, to which are attached the transoms and the rearmost left corner part of the stern. The stern was traditionally the rear left corner of the ship— however, in modern nautical terms the stern is considered the rear section of any ship. The sternpost may either be completely vertical or may be tilted or "raked" slightly aft. It rests on or "fays to" the ship's keel.

## STRAKE

A strake is part of the shell of the hull of a boat or ship which, in conjunction with the other strakes, keeps the sea out and the vessel afloat. It is a strip of planking in a wooden vessel or of plating in a metal one, running longitudinally along the vessel's side, bottom or the turn of the bilge, usually from one end of the vessel to the other.

The word strake also refers to:

- a straightedge used for levelling a bed of sand, or striking poured concrete or plaster level with the edges of the formwork or mould into which it has been poured. A strake used for flooring or paving work is often called a "screed".
- a shaped piece of wood used to contour or profile the shape of a mould, as for a bell.

## SPECIAL STRAKES

The strake immediately adjacent to the keel is known as the *garboard strake*. There are two, one on each side of the keel.

The longest continuous strake at the top of the side of the vessel on the main deck is called the *sheer strake*. In old vessels, a rubbing strake was built in just below a carvel sheer strake. It was much less broad but thicker than other strakes so that it projected and took any rubbing against piers or other boats when the boat was in use. In clinker boats, the rubbing strake was applied to the outside of the sheer strake. Others have no special name other than bottom strakes, bilge strakes and topside strakes. Many current pleasure craft reflect this history by providing the boat with a mechanically-attached (and therefore replaceable) rub rail at the location formerly occupied by a rubbing strake.

A stealer is a short strake employed to reduce the width of plank required where the girth of the hull increases. It is commonly employed in carvel and iron/steel shipbuilding, but there are very few, possibly only one type of, clinker craft that use them.

## CONSTRUCTION

Where the transverse sections of the vessel's shape are fuller, the strakes are wider. They taper toward the ends. In many larger vessels, to avoid ending with very narrow strakes, two are normally converted to one toward their ends. This one is known as a stealer or stealer strake. It was one of these which was also shaped to fit the concave line of the hull, under the counter at the sternpost. Historically, these had to be shaped by carving from an especially thick plank of wood (known as a short strake or plank). When someone was describing the epitome of thickness, he thought of the metaphor, *thick as two short planks*.

In very small boats, strakes can be made of one continuous piece of wood. Usually, they are made of more than one piece of wood or metal scarfed or butted together similar to brick wall construction. In metal vessels, the modern practice is to butt-weld them with full penetration welds.

In a riveted steel ship, the strakes were usually lapped and joggled, but where a smoother finish was sought, they too might be riveted on a butt strap, though this was weaker. In modern welded construction, the plates are normally butt-welded all round to adjoining plates within the strake and to adjoining strakes.

### Wooden Vessels

Wooden planks can be scarfed into a strake by suitably tapering the thickness of the adjoining ends. Traditionally, the adjoining faces were coated with mixed white lead and grease, then held together with copper rivets. It was important to not have the plank end as thin as a feather edge and to have the outboard thin edge at the after end of the joint. In later times, after adhesives had been developed by the aircraft industry, the joint might be made with resin glue and no rivets. In heavier carvel work, the joint would be a butt, riveted to a butt strap.

Even large clinker vessels had scarfed strakes as in this form of construction; the strakes contribute a significant proportion of the structural strength of the boat as well as keeping it water-tight.

## ASSEMBLING THE STRAKES

Two methods used to plank a wooden hull are: carvel in which the edges of the strakes butt against each other, using an internal frame structure to maintain the boat's shape and clinker in which the adjoining strakes are held together on an overlap. Their varying widths and angles of overlap determine the hull's shape. In the former form of construction, one more traditional in the Mediterranean than in north Europe produces a relatively smooth hull surface but needs caulking to make it watertight and it is relatively heavy (due to the mass of the required internal framing). Carvel boats are planked up onto pre-erected frames which become part of the boat.

The second method is clinker. It is traditional in the north and produced relatively light but strong boats. Normally, the boats are built from the garboard up so that the upper plank overlaps (aka hems) the lower on the outside. That overlap is known as the land and it is held together by copper rivets. In the twentieth century, the boat may have been built of plywood and the strakes glued together without clenched nails. Medieval boats are more likely to have been riveted with iron nails.

To rivet the lands, the planks in the two adjoining strakes are pierced, a nail inserted from the outside and a rove or rove punched over the inside end of the nail. The latter is then cut off a little proud of the rove. The nail is then clenched over the rove. When the shell of the boat is complete, the strakes are stiffened by the insertion of steam-bent timbers. These too, are riveted to the planking through the lands. The timbers, which are sometimes miscalled ribs, spread the load on the strakes and tie them together reducing the tendency of the relatively thin strake to split. With glued ply clinker construction, the timbers are unnecessary.

At the hood-ends of the strakes, where they approach the stem, they are let into each other with geralds (aka chases or gains). In these, the land of the lower strake is tapered to a feather edge at the end of the land where it is supported by the rebate formed by the apron. The strakes then meet the stem flush, which is also referred to as being hooded.

Steel ships may be plated as clinker-built vessels but more usually, they were built with strakes alternately in and out. The modern method is to butt-weld the strakes to each other as well as the plates within them, end to end. This leaves a smoother finish and is lighter.

## STRENGTH OF SHIPS

The strength of ships is a topic of key interest to naval architects and shipbuilders. Ships which are built too strong are heavy, slow, and cost extra money to build and operate since they weigh more, whilst ships which are built too weakly suffer from minor hull damage and in some extreme cases catastrophic failure and sinking.

### LOADS ON SHIP HULLS

The hulls of ships are subjected to a number of loads.

- Even when sitting at dockside or at anchor, the pressure of surrounding water displaced by the ship presses in on its hull.
- The weight of the hull, and of cargo and components within the ship bears down on the hull.
- Wind blows against the hull, and waves run into it.
- When a ship moves, there is additional hull drag, the force of propellors, water driven up against the bow.
- When a ship is loaded with cargo, it may have many times its own empty weight of cargo pushing down on the structure.

If the ship's structure, equipment, and cargo are distributed unevenly there may be large point loads into the structure, and if they are distributed differently than the distribution of buoyancy from displaced water then there are bending forces on the hull. When ships are drydocked, and when they are being built, they are supported on regularly spaced posts on their bottoms.

#### Primary hull loads, strength, and bending

The primary strength, loads, and bending of a ship's hull are the loads that affect the whole hull, viewed from front to back and top to bottom. Though this could be considered to include overall transverse loads (from side to side within the ship), generally it is applied to longitudinal loads (from end to end) only. The hull, viewed as a single beam, can bend

1. down in the center, known as sagging
2. up in the center, known as hogging.

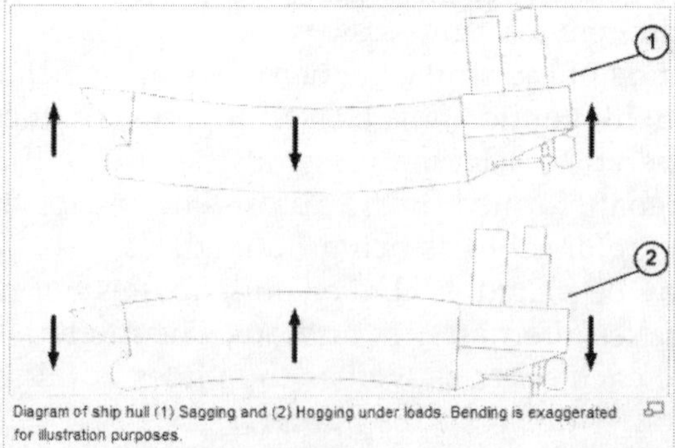


Diagram of ship hull (1) Sagging and (2) Hogging under loads. Bending is exaggerated for illustration purposes.

This can be due to:

- hull, machinery, and cargo loads
- wave loads, with the worst cases of:
  - sagging, due to a wave with length equal to the ship's length, and peaks at the bow and stern and a trough amidships
  - hogging, due to a wave with length equal to the ship's length, and a peak amidships (right at the middle of the length)

Primary hull bending loads are generally highest near the middle of the ship, and usually very minor past halfway to the bow or stern.

Primary strength calculations generally consider the midships cross section of the ship.

These calculations treat the whole ships structure as a single beam, using the simplified Euler-Bernoulli beam equation to calculate the strength of the beam in longitudinal bending.

The moment of inertia (technically, second moment of area) of the hull section is calculated by finding the neutral or central axis of the beam and then totaling up the quantity $I_y = \frac{bh^3}{12} + Ad^2$ for each section of plate or girder making up the hull, with $I_y$ being the moment of inertia of that section of material, *b*being the width (horizontal dimension) of the section, *h*being the height of the section (vertical dimension), *A*being the area of the section and *d*being the vertical distance of the center of that section from the neutral axis.

Primary strength loads calculations usually total up the ships weight and buoyancy along the hull, dividing the hull into manageable lengthwise sections such as one compartment, arbitrary ten foot segments, or some such manageable subdivision.

For each loading condition, the displaced water weight or buoyancy is calculated for that hull section based on the displaced volume of water within that hull section.

The weight of the hull is similarly calculated for that length, and the weight of equipment and systems. Cargo weight is then added in to that section depending on the loading conditions being checked.

The total still water bending moment is then calculated by integrating the difference between buoyancy and total weight along the length of the ship. For a ship in motion, additional bending moment is added to that value to account for waves it may encounter.

Standard formulas for wave height and length are used, which take ship size into account. The worst possible waves are, as noted above, where either a wave crest or trough is located exactly amidships.

Those total bending loads, including still water bending moment and wave loads, are the forces that the overall hull primary beam has to be capable of withstanding.

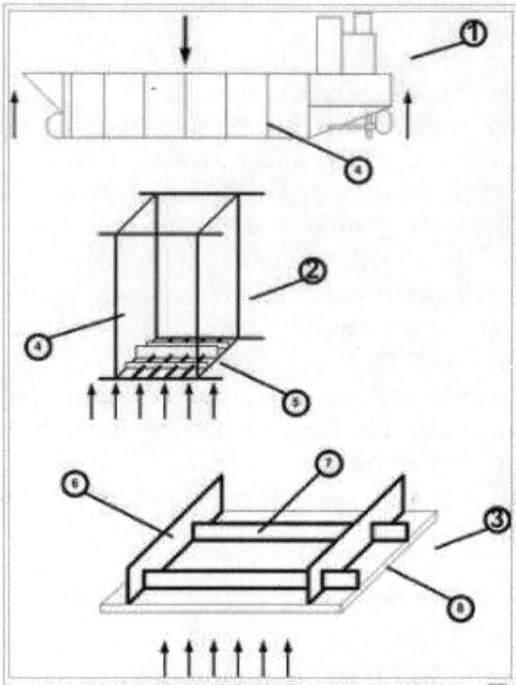

Primary (1), Secondary (2), and Tertiary (3) structural analysis of a ship hull. Depicted internal components include a watertight bulkhead (4) at the primary and secondary level, the ship's hull bottom structure including keel, keelsons, and transverse frames between two bulkheads (5) at the secondary level, and transverse frames (6), longitudinal stiffeners (7), and the hull plating (8) at the tertiary level.

**Secondary hull loads, strength, and bending**

The secondary hull loads, bending, and strength are those loads that happen to the skin structure of the ship (sides, bottom, deck) between major lengthwise subdivisions or bulkheads. For these loads, we are interested in how this shorter section behaves as an integrated beam, under the local forces of displaced water pushing back on the hull, cargo and hull and machinery weights, etc.

Unlike primary loads, secondary loads are treated as applying to a complex composite panel, supported at the sides, rather than as a simple beam. Secondary loads, strength, and bending are calculated similarly to primary loads: you determine the point and distributed loads due to displacement and weight, and determine local total forces on each unit area of the panel. Those loads then cause the composite panel to deform, usually bending inwards between bulkheads as most loads are compressive and directed inwards. Stress in the structure is calculated from the loads and bending.

**Tertiary hull loads, strength, and bending**

Tertiary strength and loads are the forces, strength, and bending response of individual sections of hull plate between stiffeners, and the behaviour of individual stiffener sections. Usually the tertiary loading is simpler to calculate: for most sections, there is a simple, maximum hydrostatic load or hydrostatic plus slamming load to calculate. The plate is supported against those loads at its edges by stiffeners and beams. The deflection of the plate (or stiffener), and additional stresses, are simply calculated from those loads and the theory of plates and shells.

**Ship hull structure elements**

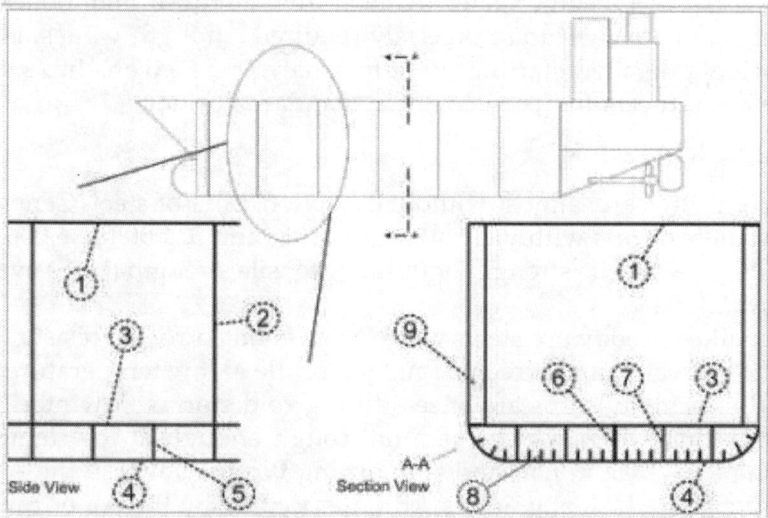


This diagram shows the key structural elements of a ship's main hull (excluding the bow, stern, and deckhouse).

1. Deck plating (a.k.a. Main Deck, Weatherdeck or Strength Deck)
2. Transverse bulkhead
3. Inner bottom shell plating
4. Hull bottom shell plating
5. Transverse frame (1 of 2)
6. Keel frame
7. Keelson (longitudinal girder) (1 of 4)
8. Longitudinal stiffener (1 of 18)
9. Hull side beam

The depicted hull is a sample small double bottom (but not double hull) oil tanker.

**TOTAL LOADS, BENDING, AND STRENGTH**

The total load on a particular section of a ship's hull is the sum total of all primary, secondary, and tertiary loads imposed on it from all factors. The typical test case for quick calculations is the middle of a hull bottom plate section between stiffeners, close to or at the midsection of the ship, somewhere midways between the keel and the side of the ship.

## STANDARD RULES

Ship classification societies such as Det Norske Veritas, American Bureau of Shipping, and Lloyd's Register have established standard calculation forms for hull loads, strength requirements, the thickness of hull plating and reinforcing stiffeners, girders, and other structures.

These methods often give a quick and dirty way to estimate strength requirements for any given ship. Almost always those methods will give conservative, or stronger than precisely required, strength values. However, they provide a detailed starting point for analyzing a given ship's structure and whether it meets industry common standards or not.

## MATERIAL RESPONSE

Modern ships are, almost without exception, built of steel. Generally this is fairly standard steel with yield strength of around 32,000 to 36,000 psi (220 to 250 MPa), and tensile strength or ultimate tensile strength (UTS) over 50,000 psi (340 MPa).

Shipbuilders today use steels which have good corrosion resistance when exposed to seawater, and which do not get brittle at low temperatures (below freezing) since many ships are at sea during cold storms in wintertime, and some older ship steels which were not tough enough at low temperature caused ships to crack in half and sink during World War II in the Atlantic.

The benchmark steel grade is ABS A, specified by the American Bureau of Shipping. This steel has a yield strength of at least 34,000 psi (230 MPa), ultimate tensile strength of 58,000 to 71,000 psi (400 to 490 MPa), must elongate at least 19% in an 8-inch (200 mm) long specimen before fracturing and 22% in a 2-inch (50 mm) long specimen.

A safety factor above the yield strength has to be applied, since steel regularly pushed to its yield strength will suffer from metal fatigue. Steels typically have a fatigue limit, below which any quantity of stress load cycles will not cause metal fatigue and cracks/failures. Ship design criteria generally assume that all normal loads on the ship, times a moderate safety factor, should be below the fatigue limit for the steel used in their construction. It is wise to assume that the ship will regularly operate fully loaded, in heavy weather and strong waves, and that it will encounter its maximum normal design operating conditions many times over its lifetime.

Designing underneath the fatigue limit coincidentally and beneficially gives large (factor of up to 6 or more) total safety factors from normal maximum operating loads to ultimate tensile failure of the structure.

But those large ultimate safety margins are not the intent: the intent is that the basic operational stress and strain on the ship, throughout its intended service life, should not cause serious fatigue cracks in the structure.

Very few ships ever see ultimate load conditions anywhere near their gross failure limits. It is likely that, without fatigue concerns, ship strength requirements would be somewhat lower.

## NUMERICAL MODELING

While it is possible to develop fairly accurate analyses of ship loads and responses by hand, or using minimal computer help such as spreadsheets, modern CAD computer programs are usually used today to generate much more detailed and powerful computer models of the structure. Finite element analysis tools are used to measure the behaviour in detail as loads are applied. These programs can handle much more complex bending and point load calculations than human engineers are able to do in reasonable amounts of time.

However, it is still important to be able to manually calculate rough behaviour of ship hulls. Engineers do not trust the output of computer programs without some general reality checking that the results are within the expected order of magnitude. And preliminary designs may be started before enough information on a structure is available to perform a computer analysis.

## STRIP-BUILT

Strip-built is a method of boat building commonly used for canoes and kayaks, but also suitable for larger boats. The process involves securing narrow, flexible strips of wood edge-to-edge around temporary forms. These are the most popular among homebuilders. Some professional builders also offered both kits and finished boats. The canoes are constructed by gluing together 1/4" x 3/4" strips of wood over a building jig consisting of station molds that define the shape of the hull. The strips may be square cut, or for a better fit, they are shaped with bead and cove router bits.

Once the strips are glued together, the inside and outside are sanded fair, and a fiberglass and epoxy covering is applied to the canoe inside and out. The fiberglass covering is transparent, allowing the wood strips to be seen. The strips are usually cedar, though sometimes pine is also used. Walnut or other contrasting woods are sometimes used as accent strips. The forms are cut as a series of cross-sections of the final design and set up along a "strong back" or other solid base. Stripping begins at the gunwale and finishes with "the football".

The strips are edge-glued to each other, being held in place with nails, staples, or simply clamped to the forms. When the glue has dried, the nails/ staples are removed and the rough hull is sanded smooth. It is then covered with a resin/epoxy impregnated overlay of fiberglass cloth, which is sanded and finished before removing the hull from the forms. The inside is then smoothed and similarly reinforced before seats, thwarts, and gunwales, are added to complete the boat.

This process is similarly suited to building kayaks. The strips are stapled onto a number of chipboard forms, the structures is sanded and fibreglassed, then the forms are removed and the hull is attached. In the 1950s, this process for building canoes was adapted from ship/boat building techniques, and refined by a group of Minnesota canoe racers; primarily Eugene Jensen, Irwin C.(Buzzy) Peterson, and Karl Ketter.

## SUPERSTRUCTURE

A superstructure is an upward extension of an existing structure above a baseline. This term is applied to various kinds of physical structures such as buildings, bridges, or ships having the degree of freedom zero [in the terms of theory of machines]. The word "superstructure" is a combination of the Latin prefix, *super*, (meaning *above, in addition*) with the Latin stem word, *structure*, (meaning *to build* or *to heap up*). In order to improve the response during earthquakes of buildings and bridges, the superstructure might be separated from its foundation by various civil engineering mechanisms or machinery. All together, these implement the system of earthquake protection called base isolation.

### ABOARD SHIPS AND LARGE BOATS

As stated above, superstructure consists of the parts of the ship or a boat, including sailboats, fishing boats, passenger ships, and submarines, that project above her main deck.

This does not usually include its masts or any armament turrets. Note that in modern times, turrets do not always carry naval artillery, but they can also carry missile launchers and/or antisubmarine warfare weapons.

The size of a watercraft's superstructure can have many implications in the performance of ships and boats, since these structures can alter their structural rigidity, their displacements, or both. These can be detrimental to any vessel's performance if they are taken into consideration incorrectly.

The height and the weight of superstructure on board a ship or a boat also affects the amount of freeboard that such a vessel requires along its sides, down to her waterline. In broad terms, the more and heavier superstructure that a ship possesses (as a fraction of her length), the less the freeboard that is needed.

### BRIDGES

On a bridge, the portion of the structure that is the span and directly receives the live load is referred to as the superstructure. In contrast, the abutment, piers, and other support structures are called the substructure.

## SEA TRIAL

A sea trial is the testing phase of a watercraft (including boats, ships, and submarines). It is also referred to as a "shakedown cruise" by many naval personnel. It is usually the last phase of construction and takes place on open water, and it can last from a few hours to many days. Sea trials are conducted to measure a vessel's performance and general seaworthiness. Testing of a vessel's speed, maneuverability, equipment and safety features are usually conducted.

Usually in attendance are technical representatives from the builder, governing and certification officials, and representatives of the owners. Successful sea trials subsequently lead to a vessel's certification for commissioning and acceptance by its owner. Although sea trials are commonly thought to be conducted only on new-built vessels (referred by shipbuilders as 'builders trials'), they are regularly conducted on commissioned vessels as well. In new vessels, they are used to determine conformance to construction specifications.

On commissioned vessels, they are generally used to confirm the impact of any modifications. Sea trials can also refer to a short test trip undertaken by a prospective buyer of a new or used vessel as one determining factor in whether to purchase the vessel.

### TYPICAL TRIALS

Sea trials are fairly standardized using technical bulletins published by ITTC, SNAME, BMT, regulatory agencies or the owners. They involve demonstrations and tests of the ship's systems and performance.

- Speed trial: The vessel is ballasted or loaded to a predetermined draft and the propulsion machinery is set to the contracted maximum service setting, usually some percentage of the machinery's maximum continuous rating. (ex: 90% MCR) The ship's heading is adjusted to have the wind and tide as close to bow-on as possible. The vessel is allowed to come to speed and the speed is continuously recorded using differential GPS. The ship is then turned through 180° and the procedure is followed again. This reduces the impact of wind and tide. The final "Trials Speed" is determined by averaging all of the measured speeds during each of the runs. This process may be repeated in various sea states.
- Crash stop: The vessel is ballasted or loaded to a predetermined draft and the propulsion machinery is set to the contracted maximum service setting, usually some percentage of the machinery's maximum continuous rating. The trial begins once the order to

"Execute Crash Stop" is given. At this point the propulsion machinery is set to full-astern and the helm is put hard-over to either port or starboard. The speed, position and heading are continuously recorded using differential GPS. The final time to stop (i.e.: ship speed is 0 knots) track line, drift (distance traveled perpendicular to the original course) and advance (distance traveled along the original course line) are all calculated. The trial may be repeated at various starting speeds.

- Endurance: The vessel is ballasted or loaded to a predetermined draft and the propulsion machinery is set to the contracted maximum service setting, usually some percentage of the machinery's maximum continuous rating. The fuel flow, exhaust and cooling water temperatures and ship's speed are all recorded.
- Maneuvering trials: A number of trials to determine the maneuverability and directional stability of the ship may be conducted. These include a direct and reverse spiral manoeuvres, zig-zag, and lateral thruster use.
- Seakeeping: Was used exclusively for passenger ships but now used in a variety of vessels. Involves measurements of ship motions in various sea states followed by a series of analyses to determine comfort levels, likelihood of sea sickness and hull damage. Trials are usually protracted in nature due to the unpredictability of finding the correct sea state and the need to conduct the trials at various headings and speeds.

## NOTEWORTHY SEA TRIALS

- RMS *Lusitania* - While steaming at high speeds, severe vibration was noted at the stern during her sea trials. This prompted her builder, John Brown & Company, to reinforce that area before acceptance by Cunard.
- SS *Normandie* - During sea trials, vibration was noted at the ship's stern. The stern was reinforced, accepted by her owners Compagnie Générale Transatlantique, and continued onto her maiden voyage. The vibration was severe enough to necessitate relocating Tourist Class passengers and some crew members with cabins near the affected area. The problem was subsequently resolved by changing her propellers to four-bladed ones from the original three-bladed ones.
- RMS *Queen Mary 2* - Her trials were conducted over two periods, September 25–29, 2003 and November 7–11, 2003, each lasting four days at sea, shuttling between the islands of Belle-Ile and L'ile d'Yeu off the French coast. On board for each set of trials were 450 people, including engineers, technicians, owner and insurance company representatives, and crew.

- USS *Thresher* (SSN-593) - Lost during deep sea diving tests on April 10, 1963.

## SEACOCK

A seacock is a valve on the hull of a boat or a ship, permitting water to flow into the boat, such as for cooling an engine or for a salt water faucet; or out of the boat, such as for a sink drain or a toilet. Seacocks are left open or are closed depending on the situation.

Seacocks feeding into or out of a closed system, like the engine cooling system, are almost always left open. Seacocks connected to something open, such as a sink drain, might be opened up in port but closed when at sea. The reason for this is that when the boat is level, the drain or other opening will always be above the waterline, and so water will only flow out. At sea, when the boat rolls in the waves, the opening may sometimes be below the waterline. If the seacock is open, water may flood the boat.

## SHIP CRADLE

A Ship cradle is a support, made of wood or metal, to hold a ship or boat upright on land so that the vessel can be built or repaired. The vessel is chocked up by wooden chocks and fixed on the cradle. The ship cradle is built near the seashore, river side or water side or in a dry dock.Sometimes the whole support is called a "cradle", sometimes each section of this support is called a cradle.

The cradle can be standing on the dock floor or it can be equipped with wheels so that the ship standing upon it can be transported from the dry docking installation to a parking area. For smaller boats, the cradle is equipped with rubber-tyred wheels, for larger and more heavy ships, the cradle is equipped with steel rail wheels.

Most cradles with steel wheels can only be transported in the longitudinal direction of the cradle. The empty cradle that is shown here is equipped with wheels that can be rotated in the transverse direction, so that the ship can be transported also over rails in the transverse direction. The fishing boat in the picture can also be transported in the longitudinal as well as the transverse direction.

In this case a transverse carriage is used between the cradle and the rails on the shipyard.

## SHIP FLOODABILITY

Floodability is a characteristic of the construction of a ship to resist flooding. Floodability is reduced by dividing the volume of the hull into watertight compartments with decks and bulkheads (which also increase the strength of ships), use of double bottom, and by other means. If a ship's hull is divided into watertight compartments, any flooding resulting from a breach of the hull can be contained in the compartments where the flooding occurs. As long as the flooding is localised, this can allow a ship to retain sufficient buoyancy to remain afloat.

The Song Dynasty Chinese author Zhu Yu wrote of watertight compartments in his book, *Pingzhou Table Talks*, written from AD 1111 to 1117 and published in 1119. Watertight compartments were frequently implemented in Asian ships, and had been implemented in the warships of Kubla Khan.

Chinese seagoing junks often had 14 crosswalls, some of which could be flooded to increase stability or for the carriage of liquids. Alexei Krylov and Stepan Makarov worked extensively on ship floodability in the early 20th century.

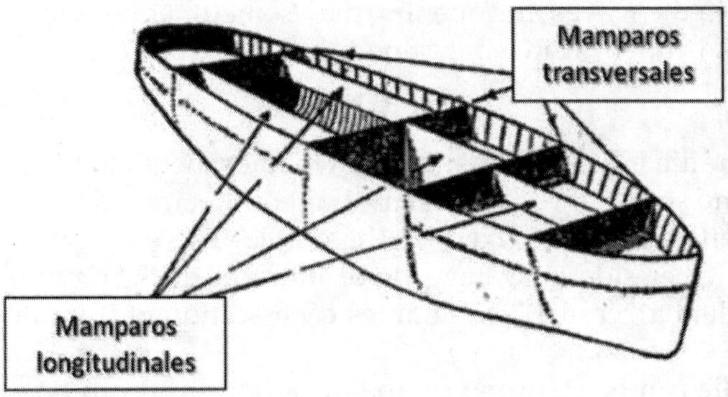


## SHIP STABILITY

Ship stability is an area of naval architecture and ship design that deals with how a ship behaves at sea, both in still water and in waves. Stability calculations focus on the center of gravity and center of buoyancy of vessels and on how these interact.

Ship stability, as it pertains to naval architecture, has existed for hundreds of years. Historically, ship stability calculations for ships relied on rule of thumb calculations, often tied to a specific system of measurement. Some of these very old equations continue to be used in naval architecture books today, however the advent of the ship model basin allows much more complex analysis.

Master shipbuilders of the past used a system of adaptive and variant design. Ships were often copied from one generation to the next with only minor changes being made, and by doing this, serious problems were not often encountered.

Ships today still use the process of adaptation and variation that has been used for hundreds of years, however computational fluid dynamics, ship model testing and a better overall understanding of fluid and ship motions has allowed much more in-depth analysis. Transverse and longitudinal waterproof bulkheads were introduced in ironclad designs between 1860 and the 1880s, anti-collision bulkheads having been made compulsory in British steam merchant ships prior to 1860. Prior to this, a hull breach in any part of a vessel could flood the entire length of the ship.

Transverse bulkheads, while expensive, increase the likelihood of ship survival in the event of damage to the hull by limiting flooding to breached compartments separated by bulkheads from undamaged ones. Longitudinal bulkheads have a similar purpose, but damaged stability effects must be taken into account to eliminate excessive heeling.

Today, most ships have means to equalize the water in sections port and starboard (cross flooding), which helps to limit the stresses experienced by the structure and also to alter the heel and/or trim of the ship.

## ADD-ON STABILITY SYSTEMS

These systems are designed to reduce the effects of waves or wind gusts. They do not increase the stability of the vessel in a calm sea. The International Maritime Organization International Convention on Load Lines does not mention active stability systems as a method of ensuring stability. The hull must be stable without active systems.

### Passive systems

#### *Bilge keel*

A bilge keel is a long fin of metal, often in a "V" shape, welded along the length of the ship at the turn of the bilge. Bilge keels are employed in pairs (one for each side of the ship). A ship may have more than one bilge keel per side, but this is rare. Bilge keels increase the hydrodynamic resistance when a vessel rolls, thus limiting the amount of roll a vessel has to endure.

#### *Outriggers*

Outriggers may be employed on certain vessels to reduce rolling. Rolling is reduced either by the force required to submerge buoyant floats or by hydrodynamic foils. In some cases these outriggers may be of sufficient size to classify the vessel as a trimaran, however on other vessels they may simply be referred to as stabilizers.

#### *Antiroll tanks*

Antiroll Tanks are tanks within the vessel fitted with baffles intended to slow the rate of water transfer from the port side of the tank to the starboard side. The tank is designed such that a larger amount of water is trapped on the higher side of the vessel. This is intended to have an effect completely opposite to that of the free surface effect.

### *Paravanes*

Paravanes may be employed by slow-moving vessels (such as fishing vessels) to reduce roll.

### Active systems

Many vessels are fitted with active stability systems. Active stability systems are defined by the need to input energy to the system in the form of a pump, hydraulic piston, or electric actuator. These systems include stabilizer fins attached to the side of the vessel or tanks in which fluid is pumped around to counteract the motion of the vessel.

### *Stabilizer fins*

Active fin stabilizers are normally used to reduce the roll that a vessel experiences while under way or, more recently, while at rest. The fins extend beyond the hull of the vessel below the waterline and alter their angle of attack depending upon heel angle and rate-of-roll of the vessel. They operate similar to airplane ailerons. Cruise ships and yachts frequently use this type of stabilizer system.

When fins are not retractable, they constitute fixed appendages to the hull, possibly extending the beam or draft envelope, requiring attention for additional hull clearances. While the typical "active fin" stabilizer will effectively counteract roll for ships under way, some modern active fin systems have been shown capable of reducing roll motion when vessels are not under way.

Referred to as zero-speed or Stabilization at Rest, these systems work by moving fins of special design, with the requisite acceleration and impulse timing to create effective roll cancellation energy.

### *Gyroscopic internal stabilizers*

Gyroscopes were first used to control a ship's roll in the late 1920s and early 1930s for warships and then passenger liners. The most ambitious use of large gyros to control a ship's roll was on an Italian passenger liner, the SS *Conte di Savoia,* in which three large Sperry gyros were mounted in the forward part of the ship.

While it proved successful in drastically reducing roll in the westbound trips, the system had to be disconnected on the eastbound leg for safety reasons. This was because with a following sea (and the deep slow rolls this generated) the vessel tended to 'hang' with the system turned on, and the inertia it generated made it harder for the vessel to right herself from heavy rolls.

Gyro stabilizers consist of a spinning flywheel and gyroscopic precession that imposes boat-righting torque on the hull structure. The angular momentum of the gyro's flywheel is a measure of the extent to which the flywheel will continue to rotate about its axis unless acted upon by an external torque. The higher the angular momentum, the greater the resisting force of the gyro to external torque (in this case more ability to cancel boat roll).

A gyroscope has three axes: a spin axis, an input axis, and an output axis. The spin axis is the axis about which the flywheel is spinning and is vertical for a boat gyro. The input axis is the axis about which input torques are applied. For a boat, the principal input axis is the longitudinal axis of the boat since that is the axis around which the boat rolls. The principal output axis is the transverse (athwartship) axis about which the gyro rotates or precesses in reaction to an input.

When the boat rolls, the rotation acts as an input to the gyro, causing the gyro to generate rotation around its output axis such that the spin axis rotates to align itself with the input axis. This output rotation is called precession and, in the boat case, the gyro will rotate fore and aft about the output or gimbal axis.

Angular momentum is the measure of effectiveness for a gyro stabilizer, analogous to horsepower ratings on a diesel engine or kilowatts on a generator. In specifications for gyro stabilizers, the total angular momentum (moment of inertia multiplied by spin speed) is the key quantity.

In modern designs, the output axis torque can be used to control the angle of the stabilizer fins to counteract the roll of the boat so that only a small gyroscope is needed. The idea for gyro controlling a ship's fin stabilizers was first proposed in 1932 by a General Electric scientist, Dr Alexanderson. He proposed a gyro to control the current to the electric motors on the stabilizer fins, with the actuating instructions being generated by thyratron vacuum tubes.

## CALCULATED STABILITY CONDITIONS

When a hull is designed, stability calculations are performed for the intact and damaged states of the vessel. Ships are usually designed to slightly exceed the stability requirements (below), as they are usually tested for this by a classification society.

### Intact stability

Intact stability calculations are relatively straightforward and involve taking all the centers of mass of objects on the vessel and the center of buoyancy of the hull. Cargo arrangements and loadings, crane operations, and the design sea states are usually taken into account.

### Damage stability (Stability in the damaged condition)

Damage stability calculations are much more complicated than intact stability. Software utilizing numerical methods are typically employed because the areas and volumes can quickly become tedious and long to compute using other methods.

The loss of stability from flooding may be due in part to the free surface effect. Water accumulating in the hull usually drains to the bilges, lowering the centre of gravity and actually decreasing (It should read as increasing, since water will add as a bottom weight there by increasing GM) the metacentric height.

This assumes the ship remains stationary and upright. However, once the ship is inclined to any degree (a wave strikes it for example), the fluid in the bilge moves to the low side. This results in a list. Stability is also lost in flooding when, for example, an empty tank is filled with seawater. The lost buoyancy of the tank results in that section of the ship lowering into the water slightly.

This creates a list unless the tank is on the centerline of the vessel. In stability calculations, when a tank is filled, its contents are assumed to be lost and replaced by seawater. If these contents are lighter than seawater, (light oil for example) then buoyancy is lost and the section lowers slightly in the water accordingly.

For merchant vessels, and increasingly for passenger vessels, the damage stability calculations are of a probabilistic nature. That is, instead of assessing the ship for one compartment failure, a situation where two or even up to three compartments are flooded will be assessed as well. This is a concept in which the chance that a compartment is damaged is combined with the consequences for the ship, resulting in a damage stability index number that has to comply with certain regulations.

## REQUIRED STABILITY

In order to be acceptable to classification societies such as the Bureau Veritas, American Bureau of Shipping, Lloyd's Register of Ships and Det Norske Veritas, the blueprints of the ship must be provided for independent review by the classification society. Calculations must also be provided which follow a structure outlined in the regulations for the country in which the ship intends to be flagged.

For U.S. flagged vessels, blueprints and stability calculations are checked against the U.S. Code of Federal Regulations and International Convention for the Safety of Life at Sea conventions. Ships are required to be stable in the conditions to which they are designed for, in both undamaged and damaged states.

The extent of damage required to design for is included in the regulations. The assumed hole is calculated as fractions of the length and breadth of the vessel, and is to be placed in the area of the ship where it would cause the most damage to vessel stability. In addition, United States Coast Guard rules apply to vessels operating in U.S. ports and in U.S. waters. Generally these

Coast Guard rules concern a minimum metacentric height or a minimum righting moment. Because different countries may have different requirements for the minimum metacentric height, most ships are now fitted with stability computers that calculate this distance on the fly based on the cargo or crew loading. There are many commercially available computer programs used for this task.

## GROSS TONNAGE

Gross tonnage (often abbreviated as GT, G.T. or gt) is a unitless index related to a ship's overall internal volume. Gross tonnage is different from gross register tonnage. Neither gross tonnage nor gross register tonnage is a measure of the ship's displacement (mass) and should not be confused with terms such as deadweight tonnage or displacement.

Gross tonnage, along with net tonnage, was defined by *The International Convention on Tonnage Measurement of Ships, 1969*, adopted by the International Maritime Organization in 1969, and came into force on July 18, 1982. These two measurements replaced gross register tonnage (GRT) and net register tonnage (NRT). Gross tonnage is calculated based on "the moulded volume of all enclosed spaces of the ship" and is used to determine things such as a ship's manning regulations, safety rules, registration fees, and port dues, whereas the older gross register tonnage is a measure of the volume of certain enclosed spaces.

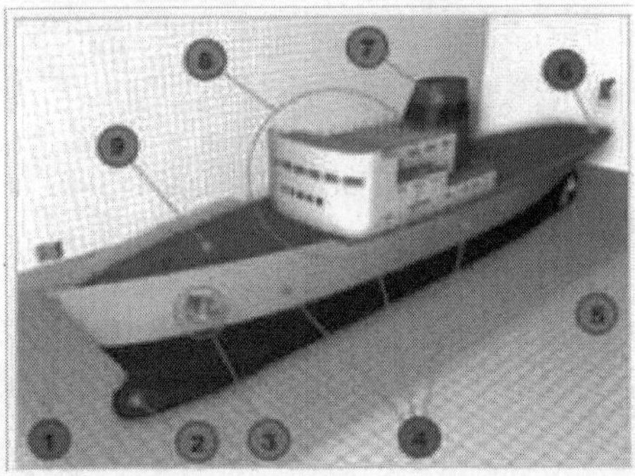

The International Convention on Tonnage Measurement of Ships, 1969 was adopted by IMO in 1969. The Convention mandated a transition from the former measurements of gross register tonnage (grt) and net register tonnage (nrt) to gross tonnage (GT) and net tonnage (NT). It was the first successful attempt to introduce a universal tonnage measurement system.

Various methods were previously used to calculate merchant ship tonnage, but they differed significantly and one single international system was needed. Previous methods traced back to George Moorsom of Great Britain's Board of Trade who devised one such method in 1854. The tonnage determination rules apply to all ships built on or after July 18, 1982. Ships built before that date were given 12 years to migrate from their existing gross register tonnage (GRT) to use of GT and NT. The phase-in period was provided to allow ships time to adjust economically, since tonnage is the basis for satisfying manning regulations and safety rules. Tonnage is also the basis for calculating registration fees and port dues. One of the Convention's goals was to ensure that the new calculated tonnages "did not differ too greatly" from the traditional gross and net register tonnages.

Both GT and NT are obtained by measuring ship's volume and then applying a mathematical formula. Gross tonnage is based on "the moulded volume of all enclosed spaces of the ship" whereas net tonnage is based on "the moulded volume of all cargo spaces of the ship". In addition, a ship's net tonnage is constrained to be no less than 30% of her gross tonnage.

**CALCULATION**

The gross tonnage calculation is defined in Regulation 3 of Annex 1 of *The International Convention on Tonnage Measurement of Ships, 1969*. It is based on two variables:

- $V$, the ship's total volume in cubic meters (m), and
- $K$, a multiplier based on the ship volume.

The value of the multiplier $K$ varies in accordance with a ship's total volume (in cubic metres) and is applied as a kind of reduction factor in determining the gross tonnage value - which does not have a unit such as cubic metres or tons. For smaller ships, $K$ is smaller, for larger ships, $K$ is larger. $K$ ranges from 0.22 to 0.32 and is calculated with a formula which uses the common or base-10 logarithm:

$K = 0.2 + 0.02 \times \log_{10}(V)$

Once $V$ and $K$ are known, gross tonnage is calculated using the formula, whereby GT is a function of V:

$GT = K \times V$

As an example, we can calculate the gross tonnage of a ship with 10,000 m total volume.

$K = 0.2 + 0.02 \times \log_{10}(V)$

$= 0.2 + 0.02 \times \log_{10})10{,}000)$

$= 0.2 + 0.02 \times 4$

$= 0.2 + 0.08$

$= 0.28$

Then the gross tonnage is calculated:

$GT = K \times V$

$= 0.28 \times 10{,}000$

$= 2{,}800$

## HAWSEHOLE

Hawsehole is a nautical term for a small hole in the hull of a ship through which *hawsers* may be passed. Also known as a cat hole. In the (British) Royal

Navy, an officer who had served as a seaman before being promoted was said to have "come in through the hawsehole."

## HAWSER

Hawser is a nautical term for a thick cable or rope used in mooring or towing a ship. A hawser passes through a hawsehole, also known as a cat hole, located on the hawse.

## HENRY GILDERSLEEVE

Henry Gildersleeve, a 19th Century American shipbuilder involved in the construction of 120 vessels of all types, was born in Gildersleeve, Connecticut (now part of Portland, Connecticut), on April 7, 1817, in the family homestead on Indian Hill Avenue.

He died of heart disease on April 9, 1894, in his residence that he built in 1853 on Main Street in Gildersleeve. At age 17, Gildersleeve left school and went to work in his father's shipyard. In 1842 he was taken into partnership with his father (Sylvester Gildersleeve) under the firm name of S. Gildersleeve & Son.

Henry Gildersleeve was part owner of many of the larger ships and schooners built by the yard and amassed a considerable fortune during the times the American merchant marine was in its glory.

### SHIP CONSTRUCTION

From 1834 to 1890, the period during which he was actively interested in the Gildersleeve shipyard, there were built 9 sloops, 46 schooners, 4 brigs, 6 barques, 11 ships, one pilot boat, one gunboat, 7 oil barges, 9 ice barges, and 26 coal, sand and cotton barges, making a total of 75 sailing vessels, 14 steamers and 31 barges, for a combined total of 120 crafts, costing $2,100,000, being an average cost of $17,500.

The most expensive sailing ships built by S. Gildersleeve & Son were the *S. Gildersleeve,* of fifteen hundred tons, built in 1854 at a cost of $59,000 and the *National Guard* of fifteen hundred tons, built in 1857 at a cost of $55,000.

### STEAM CRAFT

Henry Gildersleeve built his first propeller driven steamer in 1856. It was 275 tons and cost $20,000. He was part owner. The Gildersleeves built the United States gunboat *Cayuga* in 1861 at a cost of $125,000.

The most expensive vessel of any type built by Gildersleeve was the steamship *United States* of sixteen hundred tons, costing $150,000, built in 1864. In 1873, the steamship *City of Dallas,* costing $110,000, was built for the Mallory Line running from New York to Galveston, Texas. In 1863, he built his first steamship, *America,* 900 tons, costing $85,000, in which he was also a part owner.

## HULL (WATERCRAFT)

A hull is the watertight body of a ship or boat. Above the hull is the superstructure and/or deckhouse, where present. The line where the hull meets the water surface is called the waterline. The structure of the hull varies depending on the vessel type. In a typical modern steel ship, the structure consists of watertight and non-tight decks, major transverse and watertight (and also sometimes non-tight or longitudinal) members called bulkheads, intermediate members such as girders, stringers and webs, and minor members called ordinary transverse frames, frames, or longitudinals, depending on the structural arrangement.

The uppermost continuous deck may be called the "upper deck", "weather deck", "spar deck", "main deck", or simply "deck". The particular name given depends on the context—the type of ship or boat, the arrangement, or even where it sails. Not all hulls are decked (for instance a dinghy). In a typical wooden sailboat, the hull is constructed of wooden planking, supported by transverse frames (often referred to as ribs) and bulkheads, which are further tied together by longitudinal stringers or ceiling.

Often but not always there is a centerline longitudinal member called a keel. In fiberglass or composite hulls, the structure may resemble wooden or steel vessels to some extent, or be of a monocoque arrangement. In many cases, composite hulls are built by sandwiching thin fiber-reinforced skins over a lightweight but reasonably rigid core of foam, balsa wood, impregnated paper honeycomb or other material.

### GENERAL FEATURES

The shape of the hull is entirely dependent upon the needs of the design. Shapes range from a nearly perfect box in the case of scow barges, to a needle-sharp surface of revolution in the case of a racing multihull sailboat. The shape is chosen to strike a balance between cost, hydrostatic considerations (accommodation, load carrying and stability), hydrodynamics (speed, power requirements, and motion and behavior in a seaway) and special considerations for the ship's role, such as the rounded bow of an icebreaker or the flat bottom of a landing craft.

### HULL SHAPES

Hulls come in many varieties and can have composite shape, (e.g., a fine entry forward and inverted bell shape aft), but are grouped primarily as follows:

- Moulded, round bilged or soft-chined. Examples are the round bilge, semi-round bilge and s-bottom hull.
  - defined as smooth curves
- Chined and Hard-chined. Examples are the flat-bottom (chined), v-bottom and multi-bottom hull (hard chined).

— have at least one pronounced knuckle throughout all or most of their length

## Categorisation

After this they can be categorized as:

- Displacement the hull is supported exclusively or predominantly by *buoyancy*. Vessels that have this type of hull travel through the water at a limited rate which is defined by the waterline length. They are often heavier than planing types, though not always.
- Semi-displacement, or semi-planing the hull form is capable of developing a moderate amount of dynamic lift, however, most of the vessel's weight is still supported through buoyancy
- Planing the planing hull form is configured to develop positive dynamic pressure so that its draft decreases with increasing speed. The dynamic lift reduces the wetted surface and therefore also the drag. They are sometimes flat-bottomed, sometimes V-bottomed and sometimes round-bilged. The most common form is to have at least one chine, which makes for more efficient planing and can throw spray down. Planing hulls are more efficient at higher speeds, although they still require more energy to achieve these speeds. (see: Planing (sailing), Hull speed).

## Most used hull forms

At present, the most widely used form is the round bilge hull. The inverted bell shape of the hull, with smaller payload the waterline cross-section is less, hence the resistance is less and the speed is higher. With higher payload the outward bend provides smoother performance in waves. As such, the inverted bell shape is a popular form used with planing hulls.

## Hull forms

### *Smooth curve hulls*

Smooth curve hulls are hulls which use, just like the curved hulls, a sword or an attached keel. Semi round bilge hulls are somewhat less round. The advantage of the semi-round is that it is a nice middle between the S-bottom and chined hull.

Typical examples of a semi-round bilge hull can be found in the Centaur and Laser cruising dinghies. S-bottom hulls are hulls shaped like an *s*. In the s-bottom, the hull runs smooth to the keel.

As there are no sharp corners in the fuselage. Boats with this hull have a fixed keel, or a *kielmidzwaard*. This is a short fixed keel, with a swing keel inside. Examples of cruising dinghies that use this s-shape are the yngling and Randmeer.

### *Chined and hard-chined hulls*

A chined hull consists of straight, smooth, tall, long, or short plates, which are set at an angle to each other. The chined hull is the most simple hull shape because it works with only straight planks. These boards are often bent lengthwise.

Most home-made constructed boats are chined hull boats. The Cajun "pirogue" is an example of a craft with hard chines. Benefits of this type of boating activity is the low production cost and the (usually) fairly flat bottom, making the boat faster at planing. Chined hulls can also make use of a daggerboard or attached keel.

Chined hulls can be divided up into 3 shapes:

- V-bottom chined hulls
- Flat-bottom chined hulls
- Multi-chined hulls.

## APPENDAGES

- A protrusion below the waterline forward is called a bulbous bow and is fitted on some hulls to reduce the wave making resistance drag and thus increase fuel efficiency. Bulbs fitted at the stern are less common but accomplish a similar task.
- A keel may be fitted on a hull to increase the transverse stability, directional stability or to create lift.
- Control devices such as a rudder, trim tabs or stabilizing fins may be fitted.

## TERMS

- Bow is the front part of the hull
- Stern is the rear part of the hull
- Port is the left side of the vessel when facing the bow
- Starboard is the right side of the vessel when facing the bow
- Waterline is an imaginary line circumscribing the hull that matches the surface of the water when the hull is not moving.
- Amidships is the middle portion of the vessel in the fore and aft direction.
- Baseline is an imaginary reference line used to measure vertical distances from.

## METRICS

Hull forms are defined as follows:

- Block measures that define the principal dimensions. They are:
- Length overall (LOA) is the extreme length from one end to the other.
- Length at the waterline (LWL) is the length from the forwardmost

point of the waterline measured in profile to the stern-most point of the waterline.

- Length between perpendiculars (LBP or LPP) is the length of the summer load waterline from the stern post to the point where it crosses the stem.
- Beam or breadth (B) is the width of the hull. (ex: BWL is the maximum beam at the waterline)
- Moulded depth (D) is the vertical distance measured from the top of the keel to the underside of the upper deck at side.
- Draft (d) or (T) is the vertical distance from the bottom of the keel to the waterline.
- Freeboard (FB) is depth plus the height of the keel structure minus draft.
- Form derivatives that are calculated from the shape and the block measures. They are:
- Volume (V or Δ) is the volume of water displaced by the hull.
- Displacement (Δ) is the weight of water equivalent to the immersed volume of the hull.
- Longitudinal centre of buoyancy (LCB) is the longitudinal distance from a point of reference (often midships) to the centre of the displaced volume of water when the hull is not moving. Note that the longitudinal centre of gravity or centre of the weight of the vessel must align with the LCB when the hull is in equilibrium.
- Vertical centre of buoyancy (VCB) is the vertical distance from a point of reference (often the baseline) to the centre of the displaced volume of water when the hull is not moving.
- Longitudinal centre of floatation (LCF) is the longitudinal distance from a point of reference (often midships) to the centre of the area of waterplane when the hull is not moving. This can be visualized as being the area defined by the water's surface and the hull.
- Coefficients help compare hull forms as well:

1) Block coefficient ($C_b$) is the volume (V) divided by the LWL x BWL x T. If you draw a box around the submerged part of the ship, it is the ratio of the box volume occupied by the ship. It gives a sense of how much of the block defined by the $L_{WL}$, beam (B) & draft (T) is filled by the hull. Full forms such as oil tankers will have a high $C_b$ where fine shapes such as sailboats will have a low $C_b$.

$$C_b = \frac{V}{L_{WL} \,.\, B \,.\, T}$$

2) Midship coefficient ($C_m$ or $C_x$) is the cross-sectional area ($A_x$) of the slice at midships (or at the largest section for $C_x$) divided by beam x draft. It displays the ratio of the largest underwater section of the hull to a rectangle of the same overall width and depth as the underwater section of the hull.

This defines the fullness of the underbody. A low $C_m$ indicates a cut-away mid-section and a high $C_m$ indicates a boxy section shape. Sailboats have a cut-away mid-section with low $C_x$ whereas cargo vessels have a boxy section with high $C_x$ to help increase the $C_b$.

$$C_m = \frac{A_m}{B \, . \, T}$$

3) Prismatic coefficient ($C_p$) is the volume (V) divided by $L_{pp}$ x $A_x$. It displays the ratio of the immersed volume of the hull to a volume of a prism with equal length to the ship and cross-sectional area equal to the largest underwater section of the hull (midship section). This is used to evaluate the distribution of the volume of the underbody. A low or fine $C_p$ indicates a full mid-section and fine ends, a high or full $C_p$ indicates a boat with fuller ends. Planing hulls and other highspeed hulls tend towards a higher $C_p$. Efficient displacement hulls travelling at a low Froude number will tend to have a low $C_p$.

$$C_p = \frac{V}{L_{pp} \, . \, A_m}$$

4) Waterplane coefficient ($C_w$) is the waterplane area divided by $L_{pp}$ x B. The waterplane coefficient expresses the fullness of the waterplane, or the ratio of the waterplane area to a rectangle of the same length and width. A low $C_w$ figure indicates fine ends and a high $C_w$ figure indicates fuller ends. High $C_w$ improves stability as well as handling behavior in rough conditions.

$$C_w = \frac{A_w}{L_{pp} \, . \, B}$$

Note:

$$C_b = C_p \, . \, C_m$$

## ICE CLASS

Ships with an Ice Class have a strengthened hull to enable them to navigate through sea ice. The first requirements for merchant ships to be escorted by icebreakers were set in Finland in 1890, after winter traffic to the port of Hanko was started.

In the past, different classification organizations had different standards, rules, and ice classes. Converting between the Baltic and Arctic systems is not usually done. Recently, the two different class systems have been defined by the International Association of Classification Societies (IACS) and the International Maritime Organization (IMO).

### SIGNIFICANCE OF ICE CLASS

Not all ships are built to an ice class. Building a ship to an ice class means that the hull must be thicker, and more scantlings (aggregate of girders, beams, and bulkheads resulting in stronger structural integrity) must be in place. Sea chests (openings in the hull for seawater intake) may need to be arranged differently depending on the class. Sea bays may also be required to ensure that the sea chest does not become blocked with ice. Most of the stronger classes require several forms of rudder and propeller protection. Two rudder pintles are usually required, and strengthened propeller tips are often required in the stronger ice classes.

More watertight bulkheads, in addition to those required by a ship's normal class, are usually required. In addition, heating arrangements for fuel tanks, ballast tanks, and other tanks vital to the ship's operation may also be required depending on the class.

## ARCTIC CLASSES

The American Bureau of Shipping has a system of ice classes which includes classes A5 through A0; B0, C0, and D0. A5 class is the strongest built of the classes, with D0 being the weakest. All other major classification societies have a similar system of ice classes, and converting between ice classes is relatively easy.

In most cases only the names of the classes are changed and the specifics of the Arctic class are identical. ABS Class A5 is the only Arctic Class that may act independently in extreme Arctic waters with no limitations. Other classes are subject to limitations on time of year, required escort (always with a vessel of higher ice class) and ice conditions. The ABS database includes hundreds of Arctic Ice Classed ships, including many Arctic Research Vessels, and the entire productions of certain shipyards.

## FINNISH-SWEDISH ICE CLASS

In the Finnish-Swedish ice class rules, merchant ships operating in first-year ice in the Baltic Sea are divided into six ice classes based on requirements for hull structural design, engine output and performance in ice according to the regulations issued by the Swedish Maritime Administration and the Finnish Transport Safety Agency (TraFi). International classification societies have incorporated the Finnish-Swedish ice class rules to their own rulebooks and offer equivalent ice class notations that are recognized by the Finnish and Swedish authorities.

Ships of the highest ice class, 1A Super, are designed to operate in difficult ice conditions mainly without icebreaker assistance while ships of lower ice classes 1A, 1B and 1C are assumed to rely on icebreaker assistance. In addition there are ice class 2 for steel-hulled ships with no ice strengthening that are capable of operating independently in very light ice conditions and class 3 for vessels that do not belong to any other class such as barges.

The classes can also be spelled with roman numerals. Traffic restrictions in the Baltic Sea during winter months are based on the Finnish-Swedish ice class. These restrictions, imposed by the local maritime administrations, declare the minimum requirements for ships that are given icebreaker assistance, for example "ice class 1A, 2000 DWT".

## IACS POLAR CLASSES

The International Association of Classification Societies published a set of *Unified Requirements for Polar Class Ships* to complement the IMO *Guidelines for Ships Operating in Arctic Ice Covered Waters.*

This will effectively unity the ice classes for all IACS member societies. Seven Polar Classes are proposed in the requirements, abbreviated as PC1 through PC7

## IN-WATER SURVEY

In-water Survey (referred to by various Classification Societies as IWS, BIS, etc.) is a method of surveying the underwater parts of a ship while it is still afloat instead of having to dry-dock it for examination of these areas as was conventionally done.

For cargo ships, two surveys are required within a period of five years. One of these can be an in-water survey, provided the ship is eligible for it.

1. The ship (tanker or bulk carrier) should not be more than 15 years old.
2. The bottom of the ship should not be painted black before carrying out the survey.

## JIB

A jib is a triangular staysail that sets ahead of the foremast of a sailing vessel. Its tack is fixed to the bowsprit, to the bow, or to the deck between the bowsprit and the foremost mast. Jibs and spinnakers are the two main types of headsails on a modern boat.

## MODERN YACHTS AND SMALL CRAFT

On a boat with two staysails the inner sail is called the staysail, and the outer (foremost) is called the jib. This combination of two staysails is called a *cutter rig* (or a *yankee pair*) and a boat with one mast rigged with two staysails and a mainsail is called a cutter.

On boats with only one jib, it is common for the clew of the jib to be further aft than the mast, meaning the jib and mainsail overlap. An overlapping jib is called a *genoa jib* or simply a genoa. In acting as an airfoil, a jib has the added benefit of increasing the performance of the main sail itself by reducing turbulence on the leeward side of the sail.

On cruising yachts with more than one jib, it is common for the innermost jib to be self-tacking, either by using a boom along the foot of the sail, or by cleating the jib sheet to a track, or both. On other cruising yachts, and nearly all racing sailboats, the jib needs to be worked when tacking. On these yachts, there are two sheets attached to the clew of the jib. As the yacht comes head to wind during a tack, the active sheet is released, and the other sheet (the *lazy sheet*) on the other side of the boat is pulled in. This sheet becomes the new *active sheet* until the next tack.

## TRADITIONAL VESSELS

Schooners typically have up to three jibs. The foremost one sets on the topmast forestay and is generally called the *jib topsail*, a second on the main forestay is called *the jib*, and the innermost is called *the staysail*. Actually, all three sails are both jibs and staysails in the generic sense.

A large square-rigged ship typically has four jibs (though vessels with more or fewer exist). From forward to aft, these sails are called:

- Flying jib
- Outer jib
- Inner jib
- Fore (topmast) staysail

# JUMBOISATION

Jumboisation is a technique in shipbuilding consisting of enlarging a ship by adding an entire section to it. By contrast with refitting or installation of equipment, jumboisation is a long and complex endeavour which can require a specialised shipyard.

Enlarging a ship by jumboisation allows an increase in its value without needing to purchase or build an entirely new ship. This technique has been used on cruise ships and tankers, as well as smaller vessels like sailing or fishing ships.

## METHODS

Large ships often have a long midsection with a uniform profile. In such cases, the ship is cut in two pieces and an additional section is inserted in between. This operation must be performed in a drydock. On large ships, the additional sections are typically 20 to 30 metres long, consisting of an oil tank, a cargo ship hold, or a group of cabins, depending on the type of ship. The tanker *Seawise Giant* became the largest ship in the world after her jumboisation.

Smaller ships are usually jumboised by replacing the entire bow or stern section of the ship. This is done because the shape of their hull is usually incompatible with the previous method.

# KEEL

In boats and ships, keel can refer to either of two parts: a structural element, or a hydrodynamic element. These parts overlap. As the laying down of the keel is the initial step in construction of a ship, in British and American shipbuilding traditions the construction is dated from this event.

Only the ship's launching is considered more significant in its creation. The word can also be used to refer to a complete boat, as in keelboat or Humber keel.

## STRUCTURAL KEELS

A structural keel is a beam around which the hull of a ship is built. The keel runs in the middle of the ship, from the bow to the stern, and serves as a basic foundation or spine of the structure, providing the major source of structural strength of the hull.

The keel is generally the first part of a ship's hull to be constructed, and laying the keel, or placing the keel in the cradle in which the ship will be built, is often a momentous event in a ship's construction — so much so that the event is often marked with a ceremony, and the term *lay the keel* has entered the language as a phrase meaning the beginning of any significant undertaking.

Modern ships are now largely built in a series of pre-fabricated, complete hull sections rather than being built around a single keel, so the start of the shipbuilding process is now considered to be when the first sheet of steel is cut. The keel contributes substantially to the longitudinal strength and effectively local loading caused when docking the ship. The most common type of keel is the "flat plate keel", and this is fitted in the majority of ocean-going ships and other vessels.

A form of keel found on smaller vessels is the "bar keel", which may be fitted in trawlers, tugs, and smaller ferries. Where grounding is possible, this type of keel is suitable with its massive scantlings, but there is always a problem of the increased draft with no additional cargo capacity.

If a double bottom is fitted, the keel is almost inevitably of the flat plate type, bar keels often being associated with open floors, where the plate keel may also be fitted. Duct keels are provided in the bottom of some vessels. These run from the forward engine room bulkhead to the collision bulkhead and are utilized to carry the double bottom piping.

The piping is then accessible when cargo is loaded. If a ship suffers severe structural stress — classically during a shipwreck when running aground in a heavy sea — it is possible for the keel to break or be strained to the extent that it loses structural integrity. In this case the ship is commonly said to have "broken its back".

Such a failure means that the entire structure of the ship and its machinery has been compromised and repairing such damage would require virtually re-building the ship from the ground up. A ship that has broken its back is almost certainly unsalvagable and subsequently written off by its insurers.

## HYDRODYNAMIC KEELS

### Non-sailing keels

The keel surface on the bottom of the hull gives the ship greater directional control and stability. In non-sailing hulls, the keel helps the hull to move forward, rather than slipping to the side. In traditional boat building, this is provided by the structural keel, which projects from the bottom of the hull along most or all of its length.

In modern construction the bar keel or flat-plate keel performs the same function. There are many types of fixed keels, including full keels, long keels, fin keels, winged keels, bulb keels, and bilge keels among other designs. Deep draft ships will typically have a flat bottom and employ only bilge keels, both to aid directional control and to damp rolling motions

### Sailboat keels

In sailboats, keels use the forward motion of the boat to generate lift to counteract the leeward force of the wind. The rudimentary purpose of the keel is to convert the sideways motion of the wind when it is abeam into forward motion. A secondary purpose of the keel is to provide ballast. Keels are different from centreboards and other types of foils in that keels are made of heavy materials to provide ballast to stabilize the boat. Keels may be *fixed*, or non-movable, or they may retract to allow sailing in shallower waters.

Retracting keels may pivot (a *swing keel*) or slide upwards to retract, and are usually retracted with a winch due to the weight of the ballast. Since the keel provides far more stability when lowered than when retracted (due to the greater moment arm involved), the amount of sail carried is generally reduced when sailing with the keel retracted.

Types of non-fixed keels include swing keels and canting keels. Canting keels can be found on racing yachts, such as those competing in the Volvo Ocean Race. They provide considerably more righting moment as the keel moves out to the windward-side of the boat while using less weight. The horizontal distance from the weight to the pivot is increased, which generates a larger righting moment.

## ETYMOLOGY

The word "keel" comes from Old English *çcol*, Old Norse *kjóll*, = "ship" or "keel". It has the distinction of being regarded by some scholars as the very first word in the English language recorded in writing, having been recorded by Gildas in his 6th century Latin work *De Excidio et Conquestu Britanniae*, under the spelling *cyulae* (he was referring to the three ships that the Saxons first arrived in).

*Carina* is the Latin word for "keel" and is the origin of the term careen (to clean a keel and the hull in general, often by rolling the ship on its side). An example of this use is Careening Cove, a suburb of Sydney, Australia, where careening was carried out in early colonial days.

## KEEL LAYING

Laying the keel, or laying down is the formal recognition of the start of a ship's construction. It is often marked with a ceremony attended by dignitaries from the shipbuilding company and the ultimate owners of the ship.

Keel laying is one of the four specially-celebrated events in the life of a ship; the others are launching, commissioning, and decommissioning.

In earlier times, the event recognized as the keel laying was the initial placement of the central timber making up the backbone of a vessel, called the keel. As steel ships replaced wooden ones, the central timber gave way to a central steel beam.

Modern ships are now largely built in a series of pre-fabricated, complete hull sections rather than being built around a single keel. The event recognized as the keel laying is the first joining of modular components, or the lowering of the first module into place in the building dock. It is now often called "keel authentication", and is the ceremonial beginning of the ship's life, even though fabrication of the modules may have begun months before.

## KELSON

The kelson or keelson is the member which, particularly in a wooden vessel, lies parallel with its keel but above the transverse members such as timbers, frames or in a larger vessel, floors. It is fastened to the keel partly to impart additional longitudinal stiffness to it but principally to bind the longitudinal members (keel and hog) to the transverse members (frames and floors). In an open boat it is often fastened to the keel and hog in such a way that it can be removed for maintenance.

Again, it lies above the boat's frames or timbers as they cross the hog but in this instance, its main function is frequently to provide a means of holding down the bottom boards in such a way that they can easily be removed for maintenance. The keelson of an open boat is normally arranged to be flush with the bottom boards so as to reduce the chance of the crew's tripping over it.

In an open boat or in a larger vessel, the hog is the structural member which lies immediately above the keel to which it is permanently and securely fastened so that the two form one member to which the lowest strakes (the garboard strakes) are fastened.

## A1 (SHIPPING)

In shipping, the designation A1 is a symbol used to denote quality of construction and material. In the various shipping registers ships are classed and given a rating after an official examination, and assigned a classification mark, which appears in addition to other particulars in those shipping registers after the name of the ship.

## ABS STEELS

ABS Steels are types of structural steel which are standardized by the American Bureau of Shipping for use in shipbuilding. ABS steels come many grades in *ordinary-strength* and two levels of *higher-strength* specifications. All of these steels have been engineered to be optimal long-lived shipbuilding steels.

ABS does permit the use of other steels in shipbuilding, but discourages it, and requires more detailed engineering analysis. Ordinary-strength ABS shipbuilding steel comes in a number of grades, A, B, D, E, DS, and CS. On certified steels, the plates are marked with the grade and a preceding "AB/", e.g. AB/A etc.

Yield point for all ordinary-strength ABS steels is specified as 34,000 psi (235 MPa), except for ABS A in thicknesses of greater than 1 inch (25 mm) which has yield strength of 32,000 psi (225 MPa), and cold flange rolled sections, which have yield strength of 30,000 psi (205 MPa).

Ultimate tensile strength of ordinary strength alloys is 58,000 - 71,000 psi (400-490 MPa), except for ABS A shapes and bars with 58,000 - 80,000 psi (400-550 MPa), and cold flanged sections with 55,000 - 65,000 psi (380-450 MPa). The various grades have slightly differing alloy chemical ingredients, and differing fracture toughness.

## AFTERCASTLE

Aftercastle (or sterncastle, sometimes aftcastle) is the stern structure behind the mizzenmast and above the transom on large sailing ships, such as carracks, caravels, galleons and galleasses. It usually houses the captain's cabin and perhaps additional cabins, and is crowned by the poop deck, which on man-of-wars provided a heightened platform from which to fire upon other ships; it was also a place of defense in the event of boarding. More common, but much smaller, is the forecastle.

## AKA (SAILING)

The aka of a multihull sailboat is a member of the framework that connects the hull to the ama(s) (outrigger). The term aka originated with the proa, but is also applied to modern trimarans.

The design of the aka depends on the forces it will encounter when sailing. For example, there are two modern variations of the proa, the traditional or Pacific proa, with the ama to the windward side, and the modern Atlantic proa, with the ama to the leeward.

The windward ama provides stability by placing the center of gravity far to the windward of the sail, so it is generally heavy. Ropes leading from the mast to the ama provide the force to lift the ama, so the aka must contend mainly with compressive forces, and the weight of the crew, who generally ride on a platform running between the aka. An Atlantic proa or a trimaran rely on an ama to leeward to provide stability.

The ama provides stability by moving the center of buoyancy to the leeward side, and well designed leeward amas will also provide dynamic lift to increase the stability further. The aka for a leeward ama must be designed to handle significant amounts of torque from the lift produced by the ama. Often trimarans will have a platform between the aka, so the crew can ride out on the windward side.

While this is not truly necessary from a stability point of view (the ama generally provides a tremendous amount of lift) it does reduce the drag generated by the leeward ama. The term *waka,* like the related terms aka and ama, come from the Malay and Micronesian language group terms for parts of the outrigger canoe, and *waka* can be roughly translated as canoe. A proa consists of a waka, the main canoe-like hull; an ama, the outrigger; and aka, the poles connecting the ama to the waka. The trimaran uses the same terminology, with a center *wa'a or waka, ama* on each side, and *aka* connecting the three hulls. Anti-fouling paint Anti-fouling paint or bottom paint is a specialized coating applied to the hull of a ship or boat in order to slow the growth of organisms that attach to the hull and can affect a vessel's performance and durability.

Hull coatings may have other functions in addition to their anti-fouling properties, such as acting as a barrier against corrosion on metal hulls, or improving the flow of water past the hull of a fishing vessel or high-performance racing yacht. In the Age of Sail, sailing vessels suffered severely from the growth of barnacles and weed on the hull, called "fouling." Thin sheets of copper or Muntz metal were nailed onto the hull in an attempt to prevent marine growth.

One famous example of the traditional use of metal sheathing is the clipper *Cutty Sark,* which is preserved as a museum ship in dry-dock at Greenwich in England. Marine growth affected performance (and profitability) in many ways.

- The maximum speed of a ship decreases as its hull becomes fouled with marine growth.
- Fouling hampers a ship's ability to sail upwind.
- Some marine growth, such as shipworms, would bore into the hull causing severe damage over time.
- The ship may transport harmful marine organisms to other areas.

The inventor of the anti-fouling paint was Captain Ferdinand Gravert, born in 1847 in Glückstadt (Schleswig-Holstein, now in Germany but then Danish), who sold his chemical formula in 1913 at Taltal, Chile. Captain Alex Gravert has valuable documentation about this.

## MODERN ANTI-FOULING PAINTS

In modern times, anti-fouling paints are formulated with toxic copper, organotin compounds, or other biocides—special chemicals which impede growth of barnacles, algae, and marine organisms.

"Hard" bottom paints, or "non-sloughing" bottom paints, come in several types. "Contact leaching" paints "create a porous film on the surface. Biocides are held in the pores, and released slowly." Hard bottom paints also include Teflon and silicone coatings, which are too slippery for growth to stick. SealCoat systems, which must be professionally applied, dry with small fibers sticking out from the coating surface. These small fibers move in the water, preventing bottom growth from adhering.

## ENVIRONMENTAL CONCERNS

In the 1960s and 1970s, commercial vessels commonly used bottom paints containing tributyltin (TBT), which has been banned in the International Convention on the Control of Harmful Anti-fouling Systems on Ships of the International Maritime Organization due to its serious toxic effects on marine life (such as the collapse of a French shellfish fishery).

The Port of San Diego is investigating how to reduce copper input from copper based antifouling coatings and Washington State has passed a law which may phase in a ban on copper antifouling coatings on recreational vessels beginning in January 2018. However a similar ban was rescinded in the Netherlands after the European Union's (EU) Scientific Committee on Health and Environmental Risks (SCHER) concluded The Hague had insufficiently justified the law.

In an expert opinion, the committee concluded the Netherlands government's explanation "does not provide sufficient sound scientific evidence to show that the use of copper-based antifouling paints in leisure boats presents significant environmental risk."

"Sloughing bottom paints", or "ablative" paints, are an older type of paint designed to create a hull coating which ablates (wears off) slowly, exposing a fresh layer of biocides. Scrubbing a hull with sloughing bottom paint while it is in the water releases its biocides into the environment. One way to minimize the environmental impact from hulls with sloughing bottom paint is to have them hauled out and cleaned at boatyards with a "closed loop" system. Some innovative bottom paints that do not rely on copper or tin have been introduced.

These products have been developed in response to the increasing scrutiny that copper-based ablative bottom paints have received as environmental pollutants. One brand, EPaint, which has been used by the US Coast Guard, works by producing hydrogen peroxide in the presence of light. A possible future replacement for anti-fouling paint may be slime. A mesh would cover a ship's hull beneath which a series of pores would supply the slime compound. The compound would turn into a viscous slime on contact with water and coat the mesh. The slime would constantly slough off carrying away microorganisms and barnacle larvae.

## BALLAST POND

A ballast pond was a construction found in shipyards during the age of sail. The feature was often a prominent part of the large dockyards of the Royal Navy, where large numbers of warships would be laid up 'in ordinary' during periods of peace, when the navy retained only some of its ships in active commission.

Ships in ordinary often had their stores and guns removed, as well as their masts. Consequently their ballast, often made up of varying quantities of stone, was increased to compensate. When the ships were reactivated, and fitted with stores and guns, large quantities of ballast was removed.

The ballast that was removed would be taken on barges to the ballast pond, a large square pond dug along the sides of a tidal river, and accessed by an opening to the river by which the water flowed in and out of the pond, and by which the barges entered and exited. The ballast was then dumped in the pond.

Over time the tidal action of the water flowing in and out filtered the ballast in the pond, removing any impurities. The filtered ballast could then be recovered and returned to the dockyard to replace ballast in other ships being put in ordinary, or that otherwise needed extra ballast. One example of a ballast pond (Although now used as a mini harbour) is off Torpoint, a town close to Devonport Dockyard.

## BOOM VANG

A boom vang (US) or kicking strap (UK) is a line or piston system on a sailboat used to exert downward force on the boom and thus control the shape of the sail.

The vang typically runs from the base of the mast to a point about a third of the way out the boom. Due to the great force necessary to change the height of the boom while a boat is under sail, a line based boom vang usually includes some sort of a pulley system. Hydraulic piston vangs are used on larger sailboats and controlled by manual or electric hydraulic pumps. The term kicking strap is normally shortened to "kicker" whilst sailing, e.g. "Hard on the kicker when we get to the leeward mark".

On some sailing boats, such as the 49er, rather than a set of lines pulling the boom downwards, a rigid member is used to push the boom down. By controlling leech tension, the boom vang is one of the three methods of controlling sail twist.

On small sailboats and some cruising sailboats a vang may be omitted. If a vang is not installed, then the sheet has to try to control both horizontal and vertical angles of the boom. When the boom is near the centerline, the sheet is nearly vertical, and can exert downward force on the boom. As the sheet is loosened to increase the horizontal angle of the boom and sail, the sheet becomes horizontal and exerts less downward force. A vang works with the sheet to apply the downward force on the boom at all horizontal angles, allowing the sheet to be used to control the horizontal angle of the boom effectively.

While under sail, the opposite force to the vang is supplied by the sail itself. When the sail is furled, a topping lift supplies the upward force on the boom. Some line vang systems incorporate a piston to provide the topping lift force and to damp oscillations. Hydraulic vangs can inherently act in the topping lift role.

## BRIDGE WING

A bridge wing is a narrow walkway extending outward from both sides of a pilothouse to the full width of a ship. Officers use bridge wings when docking or maneuvering in locks and narrow waterways. Many of them are equipped with a control console allowing the rudder and engines to be controlled from them.

## BULKHEAD (BARRIER)

A bulkhead is a retaining wall, such as a bulkhead within a ship or a watershed retaining wall. It may also be used in mines, to contain flooding.

Coastal bulkheads are most often referred to as seawalls, bulkheading, or riprap revetments. These manmade structures are constructed along shorelines with the purpose of controlling beach erosion. Construction materials commonly used include wood pilings, commercially developed vinyl products, large boulders stacked to form a wall, or a seawall built of concrete or another hard substance.

Coastal property owners typically seek to develop bulkheads in an attempt to slow large landslide erosion caused by wave action.

Studies over recent decades have resulted in public awareness as to potential negative effects that bulkheads may bring to beaches and the interconnected habitat areas of fish, plants, and birds. Many states have enacted laws to protect beaches to allow for future use of the beaches, as well as protect these natural habitats. As of 1971, Washington State property owners are required to obtain a permit for the construction of bulkheads that adhere to the strict regulations of the Shoreline Management Act (SMA), enacted by the state legislature.

### EFFECTS

While bulkheads may serve their purpose to slow erosion at a bluff or beachfront, they commonly cause a domino effect of change to the beach profile. The increased wave reflection caused by their presence can result in an increased re-suspension of sand in the water in front of the bulkhead. This can lead to more sand being distributed in the alongshore direction, away from the beach profile.

Due to coastal littoral drift, the sand would then instead be distributed toward the ends of the bulkheads, leaving larger gravel and sometimes bedrock in place of the once sandy beach. Since sand is a natural habitat for several species of fish to lay their eggs and is also the only surface in which eelgrass can take root, these natural processes can no longer take place in this now sand-stripped location.

The absence of eelgrass means that the spawning habitat for herring and the protection for juvenile salmon would no longer be present. This new sand-stripped habitat also encourages other species, such as kelp, to move in.

## BULKHEAD (PARTITION)

A bulkhead is an upright wall within the hull of a ship or within the fuselage of an aeroplane. Other kinds of partition elements within a ship are decks and deckheads.

Bulkhead partitions are considered to have been a feature of Chinese junks, a type of ship. Song Dynasty author Zhu Yu (fl. 12th century) wrote in his book of 1119 that the hulls of Chinese ships had a bulkhead build. The 5th-century book *Garden of Strange Things* by Liu Jingshu mentioned that a ship could allow water to enter the bottom without sinking. Archaeological evidence of bulkhead partitions has been found on a 24 m (78 ft) long Song Dynasty ship dredged from the waters off the southern coast of China in 1973, the hull of the ship divided into twelve walled compartmental sections built watertight, dated to about 1277.

Texts written by Western writers such as Marco Polo (1254–1324), Niccolò Da Conti (1395–1469), and Benjamin Franklin (1706–1790) describe the bulkhead partitions of East Asian shipbuilding. Bulkhead partitions became widespread in Western shipbuilding during the early 19th century. Benjamin Franklin wrote in a 1787 letter that "as these vessels are not to be laden with goods, their holds may without inconvenience be divided into separate apartments, after the Chinese manner, and each of these apartments caulked tight so as to keep out water." A 19th century book on shipbuilding attributes the introduction of watertight bulkheads to Charles Wye Williams, known for his steamships.

### PURPOSE

Bulkheads in a ship serve several purposes:

- increase the structural rigidity of the vessel,
- divide functional areas into rooms and

- create watertight compartments that can contain water in the case of a hull breach or other leak.
- some bulkheads and decks are fire-resistance rated to achieve compartmentalisation, a passive fire protection measure.

On an aircraft, bulkheads divide the cabin into multiple areas. On passenger aircraft a common application is for physically dividing cabins used for different classes of service (e.g. economy and business.) On combination cargo/ passenger, or "combi" aircraft, bulkhead walls are inserted to divide areas intended for passenger seating and cargo storage.

## REQUIREMENTS OF BULKHEADS

### Fire-resistance

Openings in fire-resistance rated bulkheads and decks must be firestopped to restore the fire-resistance ratings that would otherwise be compromised, if the openings were left unsealed. The authority having jurisdiction for such measures varies depending upon the flag of the ship. Merchant vessels are typically subject to the regulations and inspections of the Coast Guards of the flag country. Combat ships are subject to the regulations set out by the navy of the country that owns the ship.

### Prevention of damage from EMI and EMP

Bulkheads and decks of warships may be fully grounded (electrically) as a countermeasure against damage from EMI and EMP due to nuclear or electromagnetic bomb detonations near the ship, which could severely damage the vital electronic systems on a ship.

In the case of firestops, cable jacketing is usually removed within the seal and firestop rubber modules are internally fitted with copper shields, which contact the cables' armour in order to ground the seal.

## OTHER USES OF THE TERM

The term was later applied to other vehicles, such as railroad cars, hopper cars trams, automobiles, aircraft or spacecraft, as well as to containers, intermediate bulk containers and fuel tanks. In some of these cases bulkheads are airtight to prevent air leakage or the spread of a fire. The term may also be used for the "end walls" of bulkhead flatcars. Mechanically, a partition or panel through which connectors pass, or a connector designed to pass through a partition.

Architecture The term is frequently used to denote any boxed in beam or other downstand from a ceiling and by extension even the vertical downstand face of an area of lower ceiling beyond. This usage presumably derives from experience on boats where to maintain the structural function personnel openings through bulkheads always retain a portion of the bulkhead crossing the head of the opening. Head strikes on these downstand elements are commonplace hence in architecture any overhead downstand element comes to be referred to as a bulkhead.

## CARLING (SAILING)

In shipbuilding, carlings are two pieces of timber laid fore and aft under the deck of a ship, from one beam to another, directly over the keel. They serve as a foundation for the whole body of the ship; on these the ledges rest, whereon the planks of the deck, and other structures are fastened. The ends of the curlings are let culvertail into the beams. The great carlings are those on which the mainmast stands. There are also *carlings of the capstan*, among others. Carling knees are timbers going traversely, from the sides to the hatchway, serving to sustain the deck on both sides.

## CARVEL (BOAT BUILDING)

Carvel built or carvel planking is a method of boat building where hull planks are fastened edge to edge, gaining support from the frame and forming a smooth surface.

In contrast with clinker built hulls, where plank edges overlap, carvel construction gives a stronger hull capable of taking a variety of full-rigged sail plans, albeit one of greater weight. In addition, it enables greater length and breadth of hull and superior sail rigs because of its strong framing, and is one of the critical developments that led to the preeminence of Western European seapower during the Age of Sail and beyond.

Carvel developed through transition from the age-old Mediterranean mortise and tenon joint method to the skeleton-first hull building technique which gradually emerged in the medieval period. The first carvel-built ships were the carracks and caravels of the fourteenth and fifteenth centuries. Both were developed in Iberia and were sailed by the kingdoms of Portugal and Spain in the early trans-oceanic voyages of the Age of Exploration.

At the same time as their appearance, the centuries-long battles to expel the Muslims from Iberia were gradually swinging to the Christian side, represented by the Kingdom of Aragon and the Kingdom of Castile, which were not united into Spain until the time of Christopher Columbus. Their invention is generally credited to the Portuguese people, who first explored south along the coast of Africa searching for a trade route to the far east in order to avoid the costly middlemen of the Eastern Mediterranean civilizations who sat upon the routes of the spice trade. Spices, in the era, were expensive luxuries and were used medicinally.

### RELATIONSHIP BETWEEN CLINKER AND CARVEL

The clinker form of construction is often linked in people's minds with the Vikings, who used this method to build their famous longships from riven timber (split wood) planks. Clinker is the British term. It is known as lapstrake in North America. In general, the languages of other countries where the method was current use some version of the word clinker.

The smoother surface of a carvel boat gives the impression at first sight that it is hydrodynamically more efficient. The lands of the planking are not there to disturb the stream line. This distribution of relative efficiency between the two forms of construction is an illusion because for given hull strength, the clinker boat is lighter because it has far less heavy timber framing. It therefore displaces less water so it has less to push aside while moving. The reduced displacement could be used to make the lines finer so as to make the passage through the water easier still.

Of course, displacement was increased as cargo was loaded but still, the clinker vessel had the advantage in efficiency as the structure was less bulky; therefore, for a given internal volume, there was a smaller external one. That means that a bulkier cargo could be carried if need be, given sufficient freeboard.

Clinker built vessels are, however, not well suited to most types of sailing rigs because they lack the internal rigidity to anchor the vessel's stays against the transverse forces generated by sailing into or across the wind with either lateen or sloop rigging methods. They also lack the internal strength to support a center-board, as well as a deep keel, drastically limiting their ability to sail across or close to the wind.

Additionally, the clinker design created a vessel which could twist and flex relative to the long axis of the vessel (from bow to stern). This gave it an advantage in North Atlantic rollers so long as the vessel was small in overall displacement.

Increasing the beam, due to the light nature of the method, did not commensurately increase the vessels survivability under the torsional forces of rolling waves, and greater beam widths may have made the resultant vessels more vulnerable. Thus, the greater rigidity of carvel built construction became necessary for larger non-coastal cargo vessels, as the twisting forces grew proportional to displaced (or cargo) weight. The physics of this imposed an upper limit on the size of clinker built vessels. Later carvel-built sailing vessels exceeded the maximum size of clinker-built ships several times over. However, clinker construction remains to this day a valuable method of construction for small wooden vessels.

A number of boat building texts are available which describe the carvel planking process in detail.

**Modern carvel methods**

Traditional carvel methods leave a small gap between each plank that in the past was filled with any suitable soft flexible fibrous material sometimes combined with a thick binding substance. This caulking would gradually wear out and the hull would leak. Likewise when the boat was beached for a length of time the planks would dry and shrink so when first refloated the hull would leak badly unless recaulked – a very time-consuming and physically demanding job.

The modern variation is to use much narrower planks which are edge glued instead of caulked. With modern powersanders a much smoother hull is produced as all the small ridges between the planks can be removed. This method started to become more common in the 1960s with the more widespread availability of fully waterproof glues such as resorcinal (red glue) and then epoxy resin.

Modern waterproof glues, especially epoxy resin, have caused revolutionary changes in both carvel and clinker style construction. Whereas in traditional construction it was the nails that provided the fastening strength, now it is the glue. It has become quite common since the 1980s for both carvel and clinker construction to rely almost completely on glue for fastening. Many small boats, especially light plywood skiffs, are built without any mechanical fastenings (e.g. nails, screw bolts) at all as the glue is in fact far stronger.

## CATHEDRAL HULL

A cathedral hull is a hull shape used in modern boats, usually power-driven. It can be thought of as a kind of vestigial trimaran in which the center hull has two smaller side hulls which are so close to the main hull that there is no longer any open space. In fact a cathedral hull is a vee-bottomed boat with sponsons which extend almost as far forward as the main hull. The airspace between the hulls may be very small or nonexistent—a good example would be the classic "Boston Whaler 13" which is nearly a flat-bottomed scow at the stern.

Depending on the proportions of the side skirts and the depth of the main hull, the cathedral hull can approach the handling characteristics of either the Hickman sea sled, or the deep-vee hull. (Worth noting is that the naval architect first credited with the successful use of the deep-vee hull form, C. Raymond Hunt, also invented the Boston Whaler). The term "cathedral hull" refers to the resemblance of a section through an inverted boat to that of a medieval cathedral. The cathedral hull configuration tends to result in a very broad bow; many such boats are completely rectangular. This provides the maximum cargo or working space for a given length and beam. The hull shape is also very stable compared to a conventional v-shaped bottom, and in either light chop conditions or above 40 knots or so is faster than a flat bottom, for the same weight, length and beam.

Cathedral hulls became popular in the 1960s and 1970s, when the use of fiberglass made economical production of this hull-form possible. The undesirable aspects of the cathedral hull are greater weight and cost, pounding in rough water, and a boxy appearance. Their advantages include high waterplane area at rest (good initial stability) a dry ride in light chop, and reasonable fuel economy at planing speeds. In exposed waters this hullform has been almost totally replaced by the modified-vee and deep-vee hull forms. After a period of reduced popularity, they live on in modified form as "deck boats" which are very popular in the inland lakes and rivers, especially in the south and midwest.

## CIRCULAR ARC HULL

The circular arc hull was created by the Swedish engineer Fredrik Ljungström. In the 1930s and 40's Ljungström successfully designed and built sailboats, commonly called the Ljungstrom sailboat.

### RIBS

The frames, or ribs, of a circular arc hull all have the same radius. On Ljungström's first models, frames were a cross-section (a single arc), but he soon realized that he needed a slightly narrower profile to cope with the sea. The result was that two arcs met at the bottom so that a slightly pointed bottom was created. The frames are tilted toward the mast so that the stern and bow lean while standing straight up.

### KEEL STOCK

Ljungstrom's yachts lack a traditional keel stock. The lower plank is twice as wide. Its curvature is a circular arc with its radius approximately equal to the length of the mast.

### PLANKING

The table planks consist entirely of parallel ribs that begin and end at the rail. The bottom (longest) plank runs from stem to stern and the top two (shortest) are located amidships at the railing on the port and starboard sides of the boat.

### ADVANTAGES

The circular arc hull's advantages are mainly in flow and manufacturing/ engineering fields. Ljungstrom was an expert technical designer and knew that every surface and curvature change causes impaired flow. Constant radii worked the best. The consistent use of arcs in a hull shape provides maximum stiffness and strength, with which thinner skins can be used.

For boat builders, the circular arc hull has several advantages:

- The planks are parallel planed ribs and need not be adapted to any keel stock.
- All ribs have the same radius and can thus be glued into the same shape. This reduces the effort in jig-building. When the planks are in place, the railing line can be cut.

The completed hull has relatively small surface area for its size, this reduced wetted area also reducing skin friction. Moreover, there are no plank ends that suck water into the wood.

The circular arc hull is very well suited for so-called strip building, where the hull is built of thin, wooden slats as a distance material between fiberglass and epoxy.

**DISADVANTAGES**

The circular arc hull, as a shallow, V-shaped hull, is not ideal for a rough sea as it does not cleave the waves. However, it can have a tendency to swage. This lightweight construction has a large number of thin ribs and planks. Thus, there may be more movement within the hull than in sturdier designs.

## CLINKER (BOAT BUILDING)

Clinker built (also known as lapstrake) is a method of boat building where the edges of hull planks overlap, called a land. In craft of any size planks are also joined end to end into a strake. The technique developed in northern Europe and was successfully used by the Vikings and typical for the Hanseatic cog.

A contrasting method, where plank edges are butted smoothly seam to seam, is known as carvel construction.

Examples of clinker-built boats directly descended from those of Viking shipbuilders are the traditional round-bottomed Thames skiffs of the River Thames, and the larger, (originally) cargo-carrying Norfolk wherries of England.

**ORIGIN**

The technique of clinker developed in the Nordic (Germanic) shipbuilding tradition as distinct from the Mediterranean mortise and tenon planking technique which was introduced to the provinces of the north in the wake of Roman expansion.

Overlapping seams already appear in the 4th century BC Hjortspring boat. The oldest evidence for a clinker-built vessel, dendrochronologically dated to 190 AD, are boat fragments which were found in recent excavations at the site of the famous Nydam Boat. The Nydam Boat itself, built ca. 320 AD, is the oldest preserved clinker-built boat.

Clinker-built ships were a trademark of Nordic navigation throughout the Middle Ages, particularly of the longships of the Viking explorers and the trading cogs of the Hanseatic League.

**Planking**

In building a simple pulling boat, the keel, hog, stem, apron, deadwoods, sternpost and perhaps transom are assembled and securely set up. In normal practice, this will be the same way up as they will be in use. From the hog, the garboard, bottom, bilge, topside and sheer strakes are planked up, held together along their lands by copper rivets. At the stem and in a double-ended boat, the sternpost, geralds are formed. That is, in each case, the land of the lower strake is tapered to a feather edge at the end of the strake where it meets the stem or stern-post.

This allows the end of the strake to be screwed to the apron with the outside of the planking mutually flush at that point and flush with the stem. This means that the boat's passage through the water will not tend to lift the ends of the planking away from the stem.

Before the next plank is laid up, the face of the land on the lower strake is bevelled to suit the angle at which the next strake will lie in relation with it. This varies all along the land. Gripes are used to hold the new strake in position on the preceding one before the fastening is done.

**Timbering or framing out**

Once the shell of planking is assembled, transverse battens of oak, ash or elm, called timbers are steam-bent to fit the internal, concave side. The timbers are frequently miscalled ribs. Elm species are not durable where the boat is used frequently in fresh water. As the timbers are bent in, they are copper riveted to the shell, through the lands of the planking.

On many clinker built craft, e.g. in Scandinavia, in Thames skiffs, and larger working craft like the coble, sawn frames are used, assembled from floors and top timbers, joggled to fit the lands. Sometimes the timbers in larger craft were also joggled before being steamed in.

With the timbers all fitted, longitudinal members are bent in. The thwart risings are fastened through the timbers with its upper edge on the level of the undersides of the thwarts. Bilge keels are added to the outside of the land on which the boat would lie on a hard surface to stiffen it and protect it from wear.

A stringer is usually fitted round the inside of each bilge to strengthen it. In a small boat, this is usually arranged to serve also as a means of retaining the bottom boards. These are removable assemblies, shaped to lie over the bottom timbers and be walked upon. They spread the stresses from the crew's weight across the bottom structure.

**Longitudinals**

Inboard of the sheer strake the heavier gunwale is similarly bent in along the line of the sheer. This part of the work is finished by fitting the breast hook and quarter knees. Swivel or crutch chocks are fitted as appropriate to the gunwale, the thwarts fitted down onto the rising and held in position by knees up to the gunwale and perhaps down onto the stringer. The structure of gunwale, rising, thwart and thwart knees greatly stiffens and strengthens the shell and turns it into a boat. There are several ways of fixing the rubbing strake but in a clinker boat, it is applied to the outside of the sheer strake.

### Fittings

Finally, the fittings such as swivels or crutch plate, painter ring, stretchers, keel and stem band are fitted and fixed with screws. In a sailing dinghy, there would be more fittings such as fairleads, horse, shroud plates, mast step, toe straps and so on.

### Finishing

That more or less finishes the boatbuilder's work but the painter has yet to varnish or paint it. At stages along the way, he will have been called in to prime the timber. This is particularly so immediately before the timbering is done.

The boatbuilder will clean up the inside of the planking and the painter will prime it and probably more, partly because it is easier that way and partly so as to put some preservative on the planking behind the timbers. Similarly, it is best to have the varnishing done after the fittings are fitted but before they are shipped. Thus, the keel band will be shaped and drilled and the screw holes drilled in the wood of keel and stem then the band will be put aside while the varnishing is done.

## FASTENINGS

The planks may be fastened together in several ways:

- with copper or iron rivets consisting of a square nail and a dish shaped washer called a rove. The land is pierced, the nail knocked through from the outside, the rove punched on while the head is held up by a dolly. The nail is cut off just proud of the rove and the cut end clenched over the rove while the dolly (a small portable anvil, usually of cylindrical shape) is used to hold the nail in place. In planking up clinker work, one man can hold both dolly and clenching hammer. Although this is common where sawn frames are to be used, boats intended for steamed timbers are usually nailed but not clenched until the timbering out is complete. As timbering is a two handed job it is more efficient to leave the clenching until help is at hand then the helper dollies up, whilst the builder sits inside the hull and clenches up.
- with iron nails with the pointed nail ends protruding on the inside of the boat, bent over and back into the wood in the form of a hook. This is a cheap and cheerful technique maybe called clinching in some parts but usually, in England at least, "turnin'-em-over". It is the sort of thing which used to be found in Scandinavian-built boats but even with iron nails, on the lands, they were usually properly

clenched over roves. Nails fastening timbers were sometimes turned over, particularly where removable bottom boards were to rest on the timbers. However, it was possible to tread the bottom boards onto the clenched nails and where marks were left, gouge out recesses to accommodate the clenched nails.

- screws were used for fixing the ends of the strakes to apron and transom and in later times, knees to gunwale and thwarts, but traditionally, this last would be done with a clench bolt or a large copper nail, clenched.
- adhesive, notably epoxy. Traditionally, lands were neither glued nor was anything used to bed them. The garboard was bedded onto the hog and keel, and the ends of the strakes onto the stem and apron using a mixture of white lead and grease. During the World Wars new techniques and materials were developed by the aircraft industry. By the mid-1950s, these were well infiltrated into the boatbuilding trade. New boats in classes of racing dinghy with clinker hulls were built as glued clinker boats. The basic construction was the same but ply planking was used and the lands were glued with no fastenings, except that the ends and garboards were still screwed to apron and hog. The need to prevent the splitting of the planks was removed by the use of ply so no timbers were used. Except for a light gunwale and wide rubbing strake, the longitudinals were omitted too. A short thwart rising and knees were glued to the planking. These boats were all decked and that is how adequate stiffness was achieved. So that the liquid glue could be laid onto the land before the next plank was assembled onto it, they were built upside down.

**Fastening the centre-line structure**

In the last few years of wooden boat construction, glue and screws took over, but until the 1950s, the keel, hog, stem, apron, deadwoods, sternpost, and perhaps transom would be fastened together by bolts set in white lead and grease.

There are three kinds of bolt, of which, nowadays, the screw bolt, with its nut and washer, is by far the most common. The second type of bolt is the *pin bolt* or *cotter bolt*, which, instead of a thread, has a tapered hole forged through the end away from the head, into which a tapered pin or cotter is knocked.

The taper is in effect a straight thread. In conjunction with a washer, this draws the bolt tight, as a nut does on a screw bolt. The third type of bolt is the *clench bolt*. It has some of the features of a rivet but was usually much longer than the normal rivet; in a wooden ship, perhaps a metre or more. For a shipwright's use, it is of copper.

A head is formed by clenching it over a washer in a swage plate. It is then knocked through a hole bored through the work to be fastened. The head is held up with a dolly and the other end is formed in the same way as the head but without the swage plate. Until well into the nineteenth century, this is what held the great ships of the world together, though some may have used iron. Until the late 1950s, the centre-line assembly of British Admiralty twenty-five foot motor cutters were fastened this way.

Where suitable metal was not available, it was possible to use treenails (pronounced trennels). They were like clench bolts but made of wood, and instead of being clenched, they had a hardwood wedge knocked into each end to spread it. The surplus was then sawn off.

## RELATIONSHIP BETWEEN CLINKER AND CARVEL

The clinker form of construction is linked in people's minds with the Vikings who used this method to build their famous longships from riven timber (split wood) planks. Clinker is the most common English term for this construction in both British and American English, though in American English the method is sometimes also known as *lapstrake.*

The smoother surface of a carvel boat gives the impression at first sight that it is hydrodynamically more efficient. The lands of the planking are not there to disturb the stream line.

This distribution of relative efficiency between the two forms of construction is an illusion because for given hull strength, the clinker boat is lighter because it does not rely upon the compressive forces of the caulking and the resulting friction to bind the skin together. It therefore displaces less water so it has less to push aside while moving.

The reduced displacement could be used to make the lines finer so as to make the passage through the water easier still. Of course, displacement was increased as cargo was loaded but still, the clinker vessel had the advantage in efficiency as the structure can be less bulky. Therefore, for a given internal volume, there was a smaller external one, meaning that a bulkier cargo could be carried if need be, given sufficient freeboard.

Additionally, the clinker built method as used by the Vikings created a vessel which could twist and flex relative to the line extending length of the vessel, bow to stern.

This gave it an advantage in North Atlantic rollers so long as the vessel was small in overall displacement. Increasing the beam, due to the light nature of the method, did not commensurately increase the vessel's survivability under the torsional forces of rolling waves, and greater beam widths may have made the resultant vessels more vulnerable.

There is an upper limit to the size of clinker built vessels, which could be and was exceeded by several orders of magnitude in later large sailing vessels incorporating carvel-built construction. Clinker requires relatively wide planking stock compared to carvel, as carvel can employ stealers to reduce plank widths amidships where the girth is greatest. The need for sufficient lap to accept the clench fastenings drives towards wider planks in proportion to thickness that can be employed in carvel. In all other areas of construction including framing, deck, etc., clinker is as capable as carvel. Clinker construction remains to this day a valuable method of construction for small wooden vessels.

## COIN CEREMONY

The Coin Ceremony is an event which takes place at the keel laying, in the early stages of a ship's construction. In it, the shipbuilders place one or two coins under the keelblock of the new ship to bless the ship and as a symbol of good fortune. The coins are not normally fixed in place and are often retrieved when the ship sails out of the dry-dock, (although they are sometimes welded to the keel).

The Mast Stepping ceremony is a similar event which occurs towards the end of a ship's construction, and involves the placing of coins underneath the mast of a ship. In shipbuilding today, the coins are normally welded beneath the radar mast.

## COMPENSATED GROSS TONNAGE

Compensated Gross Tonnage (CGT) is an indicator of the amount of work that is necessary to build a given ship and is calculated by multiplying the tonnage of a ship by a coefficient, which is determined according to type and size of a particular ship.

The standard CGT system was developed in 1977 by the OECD so that inter-country shipbuilding output could be reasonably compared, as different types of ships require a greater or lesser degree of work relative to their gross tonnage.

For example, passenger ferry of a given size would require substantially more work to build than a bulk carrier of the same size due to the differing design requirements, internal structure, and required level of detail, but simply comparing the gross tonnage or deadweight of each ship would incorrectly show that they took the same amount of work. When expanded on a national scale, this difference could greatly mislead people as to the actual maritime production capacity of a given country. The formula to calculate CGT was revised by the OECD in 2007.

## COMPOSITE SHIP

The technique of composite ship construction (wooden planking over a wrought iron frame) emerged in the mid-19th century as the final stage in the evolution of fast commercial sailing ships. Construction of wrought iron hulled vessels had begun in the 1820s and was a mature technology by the time of the launch of the SS *Great Britain* in 1843. However iron hulls could not be sheathed with copper alloy (due to bimetallic corrosion) and so would become festooned with drag-inducing weed during long voyages in the tropics.

The wooden planking of a composite ship allowed the copper sheathing essential for fast ocean crossings under sail while the iron frame made the ship relatively immune from hogging and sagging, and took up less interior space than wooden framing.

The brief reign of composite clippers as the fastest mode of transport between Europe and Asia was brought to a close by the opening of the Suez Canal in 1869 and ongoing improvement in the performance of steam ships.

## CONSUTA

Consuta was a revolutionary form of construction of watertight hulls for boats and marine aircraft, comprising four veneers of mahogany planking interleaved with waterproofed calico and stitched together with copper wire. The technique was patented by Sam Saunders of Goring-on-Thames and was first used on the 1898 umpire's steam launch of the same name. Having been restored, the steam launch Consuta was returned to service on the River Thames on 15 October 2001.

After opening the S. E. Saunders boatyard at East Cowes on the Isle of Wight, the technique was further used to build the crew and engine gondolas for HMA1 Mayfly, Britain's first airship. Later, the same technique was used to construct the hull of the Sopwith Bat Boat, the early flying boat that won the Mortimer Singer prize The technique remained in use until waterproof glues became available in the 1950s.

## COPPER SHEATHING

Copper sheathing is the practice of protecting the under-water hull of a ship or boat from the corrosive effects of salt water through the use of copper plates affixed to the outside of the hull. It was pioneered and developed by the Royal Navy during the 18th century.

### DEVELOPMENT

Deterioration of the hull of a wooden ship was a significant problem during the Age of Sail. Ships' hulls were under continuous attack by shipworm and various marine weeds, all of which had some adverse effect on the ship, be it structurally, in the case of the worm, or affecting speed and handling in the case of the weeds.

The most common methods of dealing with these problems were through the use of wood, and sometimes lead, sheathing. Expendable wood sheathing effectively provided a non-structural skin to the hull for the worm to attack, and could be easily replaced in dry dock at regular intervals. Weed, however, grew rapidly and slowed ships. Lead sheathing, while more effective than wood in mitigating these problems, reacted badly with the iron bolts of the ships causing sometimes severe damage. Even older than the sheathing methods were the various graving and paying techniques. There were three main substances used: White stuff, which was a mixture of whale oil, rosin and brimstone; Black stuff, a mixture of tar and pitch; and Brown stuff, which was simply brimstone added to Black stuff. It was common practice to first apply wood sheathing and then pay it with white stuff, although black stuff was occasionally used in this way.

The use of copper sheathing was first suggested in 1708 by Charles Perry, though it was rejected by the Navy Board on grounds of high cost and perceived maintenance difficulties. The first experiments with copper sheathing were made in the late 1750s: the bottoms and sides of several ships' keels and false keels were sheathed with copper plates. In 1761 the experiment was expanded, and the 32-gun frigate HMS *Alarm* was ordered to have her entire bottom coppered, in response to the terrible condition in which she returned from service in the West Indies. HMS *Alarm* was chosen because in 1761 a letter had been sent regarding the ship's condition, saying that the worms from the waters had taken a significant toll on the ship's wooden hull.

Before the copper plates were applied the hull was covered with Soft stuff, which was simply hair, yarn and brown paper. The copper performed very well both in protecting the hull from invasion by worm and in preventing the growth of weed for, when in contact with water, the copper produced a poisonous film, composed mainly of oxychloride, that deterred these marine organisms.

Furthermore, as this film was slightly soluble it gradually washed away, leaving no way in which marine life could attach itself to the ship. However, it was soon discovered by the Admiralty that the copper bolts used to hold the plates to the hull had reacted with the iron bolts used in the construction of the ship, rendering many bolts nearly useless. In 1766, due to the poor condition of the iron bolts, *Alarm*'s copper was removed. After this experiment, and deterred by the unanticipated and not understood electrolytic reaction between the copper and iron, lead sheathing was tried again, though it was found to be unsuitable to the task, as the plates tended to fall from the hull alarmingly quickly.

In 1768 a ship named the *Dolphin* was sheathed in the same way and sailed for a few years around the world. It came back with corrosion on the iron components of the hull; they were basically irreparable and had to be replaced. In 1769 another attempt was made at coppering a ship's hull, this time on a new ship that had been constructed using bolts made from a copper alloy.

The results were far more favourable this time, but the onset and intensification of the war with America prevented the re-bolting of the Royal Navy's ships necessary to allow a full-scale coppering programme.

## HUMPHRY DAVY'S EXPERIMENTS WITH COPPER SHEATHING

In the late 18th to early 19th century, Sir Humphry Davy performed many experiments where he had various thicknesses of copper submerged on the shore and then measured how much the sea water had degraded each one. Sheets of different metals remained in the seawater for four months and then were examined.

Two harbour ships were also used in this test, one with a zinc and the other with an iron band. Both the zinc and iron became covered in a carbonate which allowed weeds, plant life and insects to attach themselves to the metal. The sheets that had cast iron or zinc were free of any attached life forms or discoloration. The main purpose of these experiments was to determine how to lessen the corrosion that the seawater caused on unprotected copper sheathing.

Unprotected copper, such as that which was not covered in another metal such as iron, would quickly go from a reddish color to a greenish color of corrosion. When the other metal was mixed in copper in ratios from 1/40 to 1/150, there was no visible sign of corrosion and minimal weight loss. When the ratio was changed to 1/200 and 1/400, there was significant corrosion and weight loss. Davy concluded that cast iron, which was the cheapest to manufacture, was the best for protection of the copper since malleable iron and zinc wore down faster.

## WIDESPREAD IMPLEMENTATION

With the American war in full swing, the Royal Navy set about coppering the bottoms of the entire fleet. This would not have happened but for the declarations of war from France (1778), Spain (1779) and the Netherlands (1780): Britain had to face her three greatest rivals, and coppering allowed the navy to stay at sea for much longer without the need for cleaning and repairs to the underwater hull, making it a very attractive, if expensive, proposition.

Fortunately the Parys Mountain copper mine on Anglesey, Wales had recently begun large-scale production that had glutted the British market with cheap copper; however the 14 tons of metal required to copper a 74-gun third-rate ship of the line still cost £1500, compared to £262 for wood. The benefits of increased speed and time at sea were deemed by the Admiralty to justify the costs involved, and in May 1779 all ships up to and including 32 guns were ordered to be coppered when next they entered dry dock. In July this order was expanded to include ships of 44 guns and fewer.

It was then decided that the entire fleet should be coppered, due to the difficulties in maintaining a mixed fleet of coppered and non-coppered ships. 82 ships of the line had been coppered by 1781, along with 14 50-gun ships, 115 frigates, and 182 unrated vessels. By the time the war ended in 1783 problems with the hull bolts were once more becoming apparent.

Finally a suitable alloy for the hull bolts was found, that of copper and zinc. At great cost, the Admiralty decided in 1786 to go ahead with the re-bolting of every ship in the navy, thus finally eliminating the bolt corrosion problem.

This process lasted several years, after which no significant changes to the coppering system were required, and copper plating remained the standard method of protecting a ship's underwater hull until the advent of modern anti-fouling paint.

## CIVILIAN USE

With its widespread adoption by the Royal Navy, many shipping owners employed the method on their merchant vessels. They were attracted by the savings made possible by copper sheathing, despite the initial outlay. As the coppering was expensive, only the better ship owners tended to invest in the method, and as a result the use of copper sheathing tended to indicate a well-found and maintained ship, which led to Lloyd's of London charging lower insurance premiums, as the vessels were better risks.

The term *copper-bottomed* continues to be used to describe a venture, plan or investment that is safe and is certain to be successful. The related *copper-fastened* (and verb form *copperfasten*) is used similarly, though with the nuance of "secured, unambiguous", rather than "trustworthy, reliable".Corrosion in ballast tanks

Corrosion in Ballast Tanks is the deterioration process where the surface of a ballast tank progresses from microblistering, to electroendosmotic blistering, and finally to cracking of the tank steel itself.

- "Effective corrosion control in segregated water ballast spaces is probably the single most important feature, next to the integrity of the initial design, in determining the ship's effective life span and structural reliability," as said by Germanischer Lloyd's Principal surveyor.

Throughout the years the merchant fleet has become increasingly aware of the importance of avoiding corrosion in ballast tanks.

## FACTORS WITH AN INFLUENCE ON CORROSION IN BALLAST TANKS

Epoxy and modified epoxy are standard coatings used to provide protective barriers to corrosion in ballast tanks. Exposed, unprotected steel will corrode much more rapidly than steel covered with this protective layer. Many ships also use sacrificial anodes or an impressed current for additional protection. Empty ballast tanks will corrode faster than areas fully immersed due to the thin - and electo conducting - moisture film covering them.

The main factors influencing the rate of corrosion are diffusion, temperature, Conductivity, type of ions, pH, and electrochemical potential.

## REGIONS OF A BALLAST TANK

Ballast tanks do not corrode uniformly throughout the tank. Each region behaves distinctively, according to it electrochemical loading. The differences can especially be seen in empty ballast tanks. The upper sections usually corrode but the lower sections will blister.

A ballast tank has three distinct sections: 1) upper, 2) mid or "boottop" area and, 3) the "double bottom" or lower wing sections. The upper regions are constantly affected by weather. This area experiences a high degree of thermal cycling and mechanical damage through vibration. This area tends to undergo anodic oxidation more rapidly than other sections and will weaken more rapidly. This ullage or headspace area contains more oxygen and thus speeds atmospheric corrosion, as evidenced by the appearance of rust scales.

In the midsection corrodes more slowly than upper or the bottom sections of the tank. Double bottoms are prone to cathodic blistering. Temperatures in this area are much lower due to the cooling of the sea. If this extremely cathodic region is placed close to an anodic source (e.g. a corroding ballast pipe), cathodic blistering may occur especially where the epoxy coating is relatively new. Mud retained in ballast water can lead to microial corrosion.

## MARINE REGULATIONS

Many maritime accidents have been caused by corrosion, and this has led to stringent regulations concerning protective coatings for ballast tanks. *The Coating Performance Standard for Ballast Tank Coatings* (PSPC), became effective in 2008. It specifies how protective coatings should be applied during vessel construction with the intention of giving a coating a 15-year service life.

Additional regulations, such as those established by the *The International Convention for the Control and Management of Ships Ballast Water & Sediments* (SBWS) sought to avoid introducing invasive species throughout the world through ship's ballast tanks. The methods used to avoid having these invasive species surviving in ballast tanks however greatly increased the rate of corrosion.

Therefore ongoing research attempts to find water treatment systems that kill invasive species, while not having a destructive effect on the ballast tank coatings. As double-hulled tankers were introduced it meant that there was more ballast tank area had to be coated and therefore a greater capital investment for ship owners. With the onset of the OPA 90 and later the amendments to MARPOL annex 1, single hull tankers (without alternative method) have basically phased out. Modern double hull tankers, with their fully "segregated ballast tanks" propose another problem. Empty tanks act as insulation from the cold sea and allow the warm cargo areas to retain their heat longer. Corrosion rates increase with differences in temperature. Consequently, the cargo side of the ballast tank corrodes more quickly than it did with single hull tankers.

## CROW'S NEST

A crow's nest is a structure in the upper part of the main mast of a ship or a structure that is used as a lookout point.

This position ensured the best view of the approaching hazards, other ships or land. It was the best device for this purpose until the invention of radar.

In the early 19th century it was simply a barrel or a basket lashed to the tallest mast. Later it became a specially designed platform with protective railing. The barrel crow's nest was invented in 1807 by the Arctic explorer William Scoresby Sr. A statue in Whitby, North Yorkshire commemorates the event.

It should not be confused with the top, the platform in the upper part of each lower mast of a square-rigged sailing ship.

The first recorded appearance of the term was in 1807, used to describe William Scoresby's barrel crows nest platform. According to a popular naval legend, the term derives from the practice of Viking sailors, who carried crows or ravens in a cage secured to the top of the mast.

In cases of poor visibility, a crow was released and the navigator plotted a course corresponding to the bird's flight path because the crow invariably headed towards the nearest land. However other naval scholars have found no evidence of the masthead crow cage and suggest the name was coined because Scoresby's lookout platform resembled a crows nest in a tree. Sometimes the term is used metaphorically in reference to topmost structures in buildings, towers, etc.

Since the crow's nest is a point far away from the ship's centre of mass, any small movement of the ship is amplified and could lead to severe seasickness, even in accustomed sailors. Therefore, being sent to the crow's nest was also considered a punishment. The Spanish curse word *carajo* is derived from the punitive usage of the crow's nest.

In classic railroad trains, the box-like structure above the caboose, the cupola, was also called the crow's nest. It served for observation of the whole train when in motion.

In hunting a crow's nest is a blind-like structure where a hunter or a pair of hunters commit themselves to stalking game. A crow's nest is not a normal type of purchaseable blind, but it is an improvised position, built by using locally discovered natural flora (tree branches, moss, snow (during winter) or sand (during summer), etc.). A crow's nest works in most environments and provides a good look-out point (hence the name) when built in an elevated position like a hillside, or top of a hill.

## DC DISTRIBUTION SYSTEM

The DC distribution system has been proposed, as a replacement for the present AC power distribution system for ships with electric propulsion. This concept represents a new way of distributing energy for low-voltage installations on ships.

It can be used for any electrical ship application up to 20 megawatts and operates at a nominal voltage of 1000V DC. The DC distribution system is simply an extension of the multiple DC links that already exist in all propulsion and thruster drives, which usually account for more than 80 percent of the electrical power consumption on electric propulsion vessels.

### BENEFITS

In addition to boosting efficiency by up to 20 percent, other benefits include space and weight savings of up to 30 percent and flexible placement of electrical equipment. This allows for significantly more cargo space and a more functional vessel layout where the electrical system is designed around the vessel functions and not vice-versa.

The efficiency improvement is mainly achieved from the system no longer being locked at a specific frequency (usually 60 Hz on ships), even though a 60 Hz power source can also be connected to the grid. This new freedom of being able to control each power source totally independently opens up numerous ways of optimizing fuel consumption. The reduced weight and footprint of the installed electrical equipment will vary depending on the ship type and application.

One comparison using the DC distribution system instead of the traditional AC system for a Platform Supply Vessel (PSV), reduced the weight of the electrical system components from 115,520 kilograms (254,680 lb) to 85,360 kilograms (188,190 lb).

### FUEL SAVINGS

The biggest potential for fuel savings lies in the ease with which energy storage devices, such as batteries or super capacitors, can be added to the system. Energy storage will help the engines level out load variations from the thrusters and other large loads.

#### Operational optimization

DC distribution system allows for new ways of thinking regarding operational optimization. The system is flexible and can combine different energy sources such as engines, turbines, and fuel cells.

This means that there is the potential to implement an energy management system that takes into account varying fuel prices and the availability of different fuels.

## CHALLENGES

Because the main AC switchboard with its AC circuit breakers and protection relays is omitted from the new design, a new protection philosophy that fulfills class requirements is needed for selectivity and equipment protection.

ABB has proposed a solution for protecting the DC distribution system using a combination of fuses and controlled turn-off semiconductor power devices. Because all energy-producing components have controllable switching devices, the fault current can be blocked much faster than is possible with traditional circuit breakers with associated protection relays.

## DEADWEIGHT TONNAGE

Deadweight tonnage (also known as deadweight abbreviated to DWT, D.W.T., d.w.t., or dwt) is a measure of how much weight a ship is carrying or can safely carry. It is the sum of the weights of cargo, fuel, fresh water, ballast water, provisions, passengers, and crew.

The term is often used to specify a ship's maximum permissible deadweight, the DWT when the ship is fully loaded so that its Plimsoll line is at the point of submersion, although it may also denote the actual DWT of a ship not loaded to capacity.

Deadweight tonnage was historically expressed in long tons but is now usually given internationally in tonnes. Deadweight tonnage is not a measure of the ship's displacement and should not be confused with gross tonnage or net tonnage (or their more archaic forms gross register tonnage or net register tonnage).

## DECK PRISM

A deck prism is a prism inserted into the deck of a ship to provide light down below. For centuries, sailing ships used deck prisms to provide a safe source of natural sunlight to illuminate areas below decks. Before electricity, light below a vessel's deck was provided by candles, oil and kerosene lamps—all dangerous aboard a wooden ship. The deck prism was a clever solution: laid flush into the deck, the glass prism refracted and dispersed natural light into the space below from a small deck opening without weakening the planks or becoming a fire hazard.

In normal usage, the prism hangs below the overhead and disperses the light sideways; the top is flat and installed flush with the deck, becoming part of the deck. A plain flat glass window would just form a single bright spot below—not very useful for general illumination—hence the prismatic shape. On colliers (coal ships), prisms were also used to keep check on the cargo hold; light from a fire would be collected by the prism and be made visible on the deck even in daylight. The names "deck light", "dead light" or "deadlight" are sometimes used, though the latter is uncommon as a reference to prisms, as more often refers to non-opening plain-glass panels.

## DORADE BOX

A dorade box (also called a dorade vent, collector box, Charlie Noble or simply a "ventilator") is a type of vent that permits the passage of air in and out of the cabin or engine room of a boat while keeping rain, spray, and sea wash out.

The basic form is a low, rectangular box fixed to the deck or cabin top, fitted with interleaving vertical baffles. The baffles alternate to be free at the floor of the box, or free at the ceiling, forming a series of chambers. A horn-shaped ventilation cowl is usually fitted facing forward to a large hole at the top of the chamber at one end of the box with another large hole opening down into the boat from the chamber at the other end.

Limber holes perforate the wall of the box at the floor of each chamber. Dorade boxes operate on the principle that air can pass relatively freely through the chambers, yet rain or sea wash will be trapped in successive chambers and drain out the small holes in the sides of the box. The principle can be applied to other forms. For example, one variant has a circular layout with the baffles as concentric rings.

### ORIGIN

The first appearance of Dorade boxes was on the Olin Stephens-designed *Dorade*, a yacht built in 1929 for ocean racing. As originally built, the Dorade's vents led directly below, but this was found to allow water below, and the vents were modified in the early 1930s.

## DOUBLE BOTTOM

A double bottom is a ship hull design and construction method where the bottom of the ship has two complete layers of watertight hull surface: one outer layer forming the normal hull of the ship, and a second inner hull which is somewhat higher in the ship, perhaps a few feet, which forms a redundant barrier to seawater in case the outer hull is damaged and leaks. The space in between the two bottoms is often used as storage tanks for fuel or ballast water, though fuel storage in the double bottom is not allowed for newbuilt ships since 2007, due to MARPOL 73/78.

Double bottoms are significantly safer than single bottoms. In case of grounding or other underwater damage, most of the time the damage is limited to flooding the bottom compartment, and the main occupied areas of the ship remain intact. For this reason, double bottoms have been required in all passenger ships for decades as part of the Safety Of Life At Sea or SOLAS Convention. An even more extensive protection is available as a double hull, where the second hull layer extends up the sides of the ship as well as in the bottom. A double bottom also conveniently forms a stiff and strong girder or beam structure with the two hull plating layers as upper and lower plates for a composite beam.

This greatly strengthens the hull in secondary hull bending and strength, and to some degree in primary hull bending and strength.

Some of the advantages of double bottom in ships are:

- provide protection to hull in case of grounding
- provide great longitudinal strength
- can be used as tanks to carry oil, ballast water or fresh water
- helps to prevent pollution in case of liquid cargo (like oil in tankers)
- helps to maintain stability of ship
- the tank top acts as a platform for machinery and cargo

## DOUBLE HULL

A double hull is a ship hull design and construction method where the bottom and sides of the ship have two complete layers of watertight hull surface: one outer layer forming the normal hull of the ship, and a second inner hull which is some distance inboard, typically by a few feet, which forms a redundant barrier to seawater in case the outer hull is damaged and leaks.

The space between the two hulls is sometimes used for storage of fuel or ballast water. Double hulls are a more extensive safety measure than double bottoms, which have two hull layers only in the bottom of the ship but not the sides.

In low-energy collisions, double hulls can prevent flooding beyond the penetrated compartment. In high-energy collisions, however, the distance to the inner hull is not sufficient and the inner compartment is penetrated as well.

Double hulls or double bottoms have been required in all passenger ships for decades as part of the Safety Of Life At Sea or SOLAS Convention. One of the disadvantages of a double hull is that the stability of a ship can be less than that of a single hull. Because the double hull raises the centre of gravity, the metacentric height is reduced.

### OIL TANKERS

Double hulls' ability to prevent or reduce oil spills led to double hulls being standardized for other types of ships including oil tankers by the International Convention for the Prevention of Pollution from Ships or MARPOL Convention. A double hull does not protect against major, high-energy collisions or groundings which cause the majority of oil pollution, despite this being the reason that the double hull was mandated by United

States legislation. After the Exxon Valdez oil spill disaster, when that ship grounded on Bligh Reef outside the port of Valdez, Alaska, the US Government required all new oil tankers built for use between US ports to be equipped with a full double hull.

However, the damage to the Exxon Valdez penetrated sections of the hull (the slops oil tanks, or slop tanks) that were protected by a double bottom, or partial double hull.

Furthermore, a double-hulled tanker does not need longitudinal bulkheads for longitudinal strength, as the inner hull already provides this. Eliminating longitudinal bulkheads would result in much wider tanks, significantly increasing the free surface effect. However, this problem is easily corrected with the addition of anti-slosh baffles and partial bulkheads.

### SUBMARINES

In submarine hulls, the double hull structure is significantly different, consisting of an outer light hull and inner pressure hull, with the outer hull intended more to provide a more hydrodynamic shape for the submarine than the cylindrical inner pressure hull.

## FITTING-OUT

Fitting-out, or "outfitting", is the process in modern shipbuilding that follows the float-out of a vessel and precedes sea trials. It is the period when all the remaining construction of the ship is completed and readied for delivery to her owners. Since most of the fitting-out process is interior work, this stage can overlap with latter stages, such as the sea trials.

### LAUNCHING/FLOATING

After a vessel has been floated (in contemporary shipbuilding) or launched (in traditional shipbuilding), it is then towed out of its drydock and moored at a fitting-out berth. While still afloat, its construction is then continued.

Depending on the type of vessel, fitting-out can last weeks or many months. Vessels with comparatively minimal space for human occupation, such as oil tankers, bulk carriers and container ships, can take the least time for fitting. Conversely, passenger ships take the longest. The process can include:

- completion of the superstructure,
- installation of the ship's power plant, engines, and other machinery,
- interior equipment and systems, including electrical, plumbing, and HVAC,
- finishing of interior spaces,
- and installation of furnishings.

Whatever construction is completed during fitting is also dependent on the shipyard's capabilities and the availability of equipment prior to floating. For example, Queen Mary 2 was originally to have its propeller pods installed prior to floating, but this was not done until the fitting-out stage. Contemporary ship construction usually has the vessel returning to drydock several times again for installation of propulsion mechanisms (such as propulsion pods in contemporary vessels) and for the painting of surfaces below the waterline.

## FLAT SPLINE

A spline or the more modern term flexible curve consists of a long strip fixed in position at a number of points that relaxes to form and hold a smooth curve passing through those points for the purpose of transferring that curve to another material.

Before computers were used for creating engineering designs, drafting tools were employed by designers drawing by hand. To draw curves, especially for shipbuilding, draftsmen often used long, thin, flexible strips of wood, plastic, or metal called splines (or laths, not to be confused with lathes). The splines were held in place with lead weights (called ducks because of their duck-like shape).

The elasticity of the spline material combined with the constraint of the control points, or knots, would cause the strip to take the shape that minimized the energy required for bending it between the fixed points, this being the smoothest possible shape. Splines are more recently referred to as flexible curves and perform much of the original function. The main difference between splines and flexible curves is that the control points of flexible curves are entirely internal in their housing.

This has one advantage over splines: whereas the draftsman had to first set up the spline and control points before moving the object to be marked to the spline in that order only, with flexible curves the draftsman can also set up the flexible curve before carefully moving the flexible curve to the object to be marked. One can recreate an original draftsman's spline device with weights and a length of thin stiff plastic or rubber tubing. The weights are attached to the tube (by gluing or pinning). The tubing is then placed over drawing paper. Crosses are marked on the paper to designate the knots or control points. The tube is then adjusted so that it passes over the control points.

In 1946, mathematicians started studying the spline shape, and derived the piecewise polynomial formula known as the spline curve or spline function. This has led to the widespread use of such functions in computer-aided design, especially in the surface designs of vehicles. I. J. Schoenberg gave the spline function its name after its resemblance to the mechanical spline used by draftsmen.

The origins of the spline in wood-working may show in the conjectured etymology which connects the word *spline* to the word *splinter*. Later craftsmen have made splines out of rubber, steel, and other elastomeric materials. Spline devices help bend the wood for pianos, violins, violas, etc. The Wright brothers used one to shape the wings of their aircraft.

## FLOAT-OUT

Float-out is the process in modern shipbuilding that follows the keel laying and precedes the fitting-out process.

It is analogous to launching a ship, a specific process that has largely been discontinued in modern shipbuilding. Both floating-out and launching are the times when the ship leaves dry land and becomes waterborne for the first time, and often take place during ceremonies celebrating and commemorating that event.

## FORECASTLE

Forecastle refers to the upper deck of a sailing ship forward of the foremast, or the forward part of a ship with the sailors' living quarters. Related to the latter meaning is the phrase "before the mast" which denotes anything related to ordinary sailors, as opposed to a ship's officers. In medieval shipbuilding, a ship of war was usually equipped with a tall, multi-deck castle-like structure in the bow of the ship.

It served as a platform for archers to shoot down on enemy ships, or as a defensive stronghold if the ship were boarded. A similar but usually much larger structure, called the aftcastle, was at the aft end of the ship, often stretching all the way from the main mast to the stern.

Having such tall upper works on the ship was detrimental to sailing performance. As cannons were introduced and gunfire replaced boarding as the primary means of naval combat during the 16th century, the medieval forecastle was no longer needed, and later ships such as the galleon had only a low, one-deck high forecastle.

By the end of the 19th Century, raised forecastle had become a typical feature on warships again, in an attempt to keep forward gun positions from getting unacceptably wet on heavy seas. In addition the forecastle may provide additional crew's quarters as in the past, and may contain essential machinery such as the anchor windlass.

A disadvantage of such a design is the structural weakness at the forecastle 'break' (the rear end of the forecastle with the main deck behind and below) relative to a flush deck structure. Some sailing ships and many modern non-sail ships have no forecastle as such at all but the name is still used to indicate the foremost part of the upper deck – although often called the *foredeck* – and for any crews quarters in the bow of the ship, even if below the main deck.

## FROUDE–KRYLOV FORCE

In fluid dynamics, the Froude–Krylov force — sometimes also called the Froude–Kriloff force — is a hydrodynamical force named after William Froude and Alexei Krylov.

The Froude–Krylov force is the force introduced by the unsteady pressure field generated by *undisturbed* waves. The Froude–Krylov force does, together with the diffraction force, make up the total non-viscous forces acting on a floating body in regular waves. The diffraction force is due to the floating body disturbing the waves.

### FORMULAS

The Froude–Krylov force can be calculated from:

$$\vec{F}_{FK} = -\iint_{Sw} p\vec{n}\, ds,$$

where

- $\vec{F}_{FK}$ is the Froude–Krylov force,
- $S_w$ is the wetted surface of the floating body,
- $p$ is the pressure in the undisturbed waves and
- $\vec{n}$ the body's normal vector pointing into the water.

In the simplest case the formula may be expressed as the product of the wetted surface area (A) of the floating body, and the dynamic pressure acting from the waves on the body:

$$F_{FK} = A \cdot pdyn \cdot$$

The dynamic pressure, $pdyn$, close to the surface, is given by:

$$p_{dyn} = \rho \cdot g \cdot H / 2$$

where

- $\rho$ is the sea water density (approx. 1030 kg/m$^3$)
- $g$ is the acceleration due to the earth's gravity (9.81 m/s$^2$)
- $H$ is the wave height from crest to trough.

## FURRING

In light-frame construction, furring strips are long thin strips of wood or metal used to make backing surfaces to support the finished surfaces in a room. Furring refers to the backing surface, the process of installing it, and may also refer to the strips themselves.

"Firring" is a U.K. term for wood strips which are usually 50 mm wide, tapered and fixed above wood roof joists to provide drainage falls below roof boarding. Furring strips themselves are typically referred to as "battens" in the U.K.

Furring strips typically measure 1" x 2" or 1" x 3". They can be laid out perpendicular to studs or joists and nailed to them, or set vertically against an existing wall surface. The spacing between the strips depends on the type of finishing material.

Wider spacing is typically used behind the heavy boards that support ceramic tiles. Closely spaced strips are needed for thin panelling or plaster. The use of strips with plaster, however, is called either lath and plaster or wattle and daub.

Metal furring strips can be used in towns where fire-proof supporting elements are required by the local building code. Often called "hat channels" to describe the profile (cross section), they consist of two flanges on each side of a trapezoid shape, 7/8" thick.

Furring is also used to support roof materials and may be seen under barn and shed roofs, but is often replaced in contemporary times by labor-saving plywood. Drywall has become the most common interior wall finishing material and doesn't need furring due to its strength. Furring is still used in remodeling work to fill out uneven sections for resurfacing, or to add room for insulation.

**SHIP REPAIR**

Furring, is a type of ship rebuilding method indicative of the late 16th century and early 17th century England. It was adopted as a remedial process to solve crank ships that were built too narrow. When a ship could not bear sail or was too narrow or her bearing laid too low, a second layer of frames were attached to the first to make her broader and lay her bearing higher. This was done by ripping the planks off and applying the second frames on top of the original frames, and then adding the planks back on.

"They commonly fur some two or three strakes under water and as much above, according as the ship requires, more or less." (Mainwaring, 153) Although this appeared to have fixed the problems with 'makeshift corrections' due to miscalculation, it was seen as a poor remedy and at times was used as a black listing method among shipwrights during the end of the 16th century and beginning of the 17th century England. (ex. Phineas Pett and the *Prince Royal*)

Because it was seen as such an incompetent occurrence to miscalculate while building, even the writers of the time had plenty to say about its uncertainties: "I think in all the world there are not so many ships furred as are in England, and it is a pity that there is no order taken either for the punishing of those who build such ships or the preventing of it, for it is an infinite loss to the owners and an utter spoiling and disgrace to all ships that are so handled (Mainwaring, 153)". The only archaeological evidence of furring found to date, is the Princes Channel Wreck, excavated and recovered by Wessex Archaeology. There has been debate as to the preliminary naming of the wreck, as it was found in the Princes Channel at the mouth of the Thames estuary. However the more popular name of Gresham Ship was introduced when the guns on board the vessel revealed a grasshopper insignia, which was the motif made by Sir Thomas Gresham, founder of the Royal Exchange.

# 4

# Ship Compartments

## ARMORED CITADEL

In a ship an armored citadel is an armored box enclosing the machinery and magazine spaces. It was formed by the armored deck, the waterline belt and the transverse bulkheads.

## SAFE ROOM

A safe room or panic room is a fortified room which is installed in a private residence or business to provide a safe shelter, or hiding place, for the inhabitants in the event of a break-in, home invasion, tornado, or other threat. Safe rooms usually contain communications equipment, so that law enforcement authorities can be contacted.

### CONSTRUCTION TECHNIQUES

The simplest safe room is simply a closet with the hollow-core door replaced with an exterior-grade solid-core door that has a deadbolt and longer hinge screws and strike-plate screws to resist battering. Sometimes, the ceiling is reinforced, or gated, to prevent easy access from the attic or from an overhead crawl space.

More expensive safe rooms, such as those constructed for celebrities and executives, have walls and a door reinforced with sheets of steel, Kevlar, or bullet-resistant fibreglass. The hinges and strike plate are often reinforced with long screws. Some safe rooms may also have externally-vented ventilation systems and a separate telephone connection. They might also connect to an escape shaft.

Safe rooms in the basement can be built with concrete walls, a building technique that is normally not possible on the upper floors of wood-framed structures unless there is substantial structural reinforcement to the building.

The U.S. State Department often uses steel grillwork much like a jail to seal off parts of a home used by U.S. Foreign Service members overseas when they are living in cities with a high crime threat. In some cities the entire upstairs area is grilled off as well as every window and door to the home.

Other homes have steel doors to one or more bedrooms that can be bolted closed to provide time for security forces to arrive.

For strong storms or tornadoes, a storm safe room must be built to withstand high winds and flying debris, even if the rest of the residence becomes severely damaged or destroyed; specific concerns:

- The safe room should be adequately anchored to the foundation to resist overturning and uplift.
- The walls, ceiling, and door of the shelter should withstand wind pressure and resist penetration by windborne objects or falling debris.
- The connections between all parts of the safe room should be strong enough to resist separation by wind.
- Sections of either interior or exterior residence walls which are used as walls of the safe room, should be separated from the structure of the residence so that damage to the residence will not cause damage to the safe room.

## FEATURES

Safe rooms may contain communications equipment, such as a cellular telephone, land-line telephone or an amateur radio transceiver, so that law enforcement authorities can be contacted. There may also be a monitor for external security cameras and an alarm system. In basic safe rooms, a peephole in the door may be used for a similar purpose. Safe rooms are typically stocked with basic emergency and survival items such as a flashlight, blankets, a first-aid kit, water, packaged food, self-defence tools, a gas mask, and a simple portable toilet.

Safe rooms can be hidden behind many household features, such as mirrors, wardrobes, bookcases and even fireplaces.

## CITADEL ON SHIPS

Warships with CBRN protection generally have a central citadel, with a degree of armour protection as well as a filtered air system. Safe rooms on civilian ships are sometimes called "citadels", are increasingly being installed as a countermeasure against piracy. When attacked, the crew can retreat into the safe room and call for help (which in the case of ships of some countries may include the intervention of military forces).

Because of the nature of ship construction, the safe room is typically constructed in a concealed location within a void within the ship, to resist efforts by the pirates to find the crew before help arrives.

Safe rooms sometimes have facilities to allow the crew to remotely disable the ship's engines and electronic systems, making it impossible for the pirates to sail the ship to a location they control. The safe room is also typically armoured against direct physical attack, to allow the crew to remain safe for a few hours, even if located by the pirates, and to allow rescuing forces full scope for the use of armed force to re-take the ship without risk to the crew.

The effect of the safe room is thus to deny the pirates access to the crew for hostage-taking, to remove the capability to move the ship to a location favourable for the pirates, and to facilitate retaking the ship by armed force without risk to the crew. The retreat of the crew to a safe room could encourage the pirates to leave the ship of their own volition.

Safe rooms have been used as a defensive measure in ships threatened by piracy in Somalia. In 2010, 4,185 seafarers had been attacked and 1,090 taken hostage, but 342 were kept safe in a citadel.

### NOTABLE USES

On 1 January 2010, a panic room was instrumental in saving Danish cartoonist Kurt Westergaard from an Islamic fundamentalist attacker who wanted to kill him for having drawn a controversial cartoon of Muhammad in 2005. The secure room prevented Westergaard from being reached by the axe-wielding intruder, who was shot and incapacitated while resisting arrest after police arrived.

During the 2011 Cairo Israeli embassy attack, rioters forcefully entered the Israeli embassy in Cairo. During the attack the eight-person Israeli diplomatic staff were kept safe and separated from the rioters by the steel door of the safe room in which they had taken refuge. They were rescued by Egyptian commandos and flown out of Egypt with their families.

## SICK BAY

A sick bay is a compartment in a ship used for medical purposes — the ship's hospital. The sick bay will contain the ship's medicine chest, which may be divided into separate cabinets, such as a refrigerator for medicines requiring cold storage and a locked cabinet for controlled substances such as morphine.

The sick bay and the medicine chest should be kept locked, with the keys only being available to the medical officer and the ship's master. The term is also applied ashore by the United States Navy and Marine Corps to treatment clinics on naval stations and Marine bases. Sick bays appear in popular science fiction franchises, such as Battlestar Galactica or the Star Trek series, as the medical facility on board a starship.

## WARDROOM

The wardroom is the mess-cabin of naval commissioned officers above the rank of Midshipman. The term *the wardroom* is also used to refer to (metonymically) those individuals with the right to occupy that wardroom, meaning "the officers of the wardroom". It provides a place of recreation as well as being a dining room. Usually, a galley or scullery adjoins the wardroom. Service is provided by stewards. There is usually a bar, where drinks can be purchased. Ships can be either "wet" or "dry": the former allowing the consumption of alcohol at sea, while the latter only allows alcohol when alongside at port, if at all. (Ships of the United States Navy have not allowed alcohol consumption onboard since 1913).

Wardrooms have rules governing etiquette. Traditionally considered taboo are three topics: politics, religion, and sex (earlier guidebooks referred to the latter as *ladies*). On large ships in peacetime, talking about professional business is frowned upon. It is also considered inappropriate to perform work or to meet with subordinates in a wardroom.

The ship's executive officer is usually the *Mess President*. The Commanding Officer is normally not a member of the wardroom, but is normally invited to join the members for special occasions.

## PILOTHOUSE

A pilothouse or pilot-house is a glass-enclosed room from which a ship is controlled by the ship's pilot. The pilothouse also is known as the wheelhouse. It usually is located on top of the largest deck (the texas deck) or highest deck. The steering wheel or helm, compass, engine order telegraph, and chart table are located here.

On many ships, especially military and cruise ships, the pilothouse is considerably larger and combined with a number of other control structures as part of the ship's bridge. Pilothouses may also be a feature of smaller vessels.

## HOLD (SHIP)

A ship's hold is a space for carrying cargo. Cargo in holds may be either packaged in crates, bales, etc., or unpackaged (bulk cargo). Access to holds is by a large hatch at the top. Ships have had holds for centuries; an alternative way to carry cargo is in standardised shipping containers, which may be loaded into appropriate holds or carried on deck.

Holds in older ships were below the orlop deck, the lower part of the interior of a ship's hull, especially when considered as storage space, as for cargo. In later merchant vessels it extended up through the decks to the underside of the weather deck.

## HEAD (WATERCRAFT)

The head (or heads) is a ship's toilet. The name derives from sailing ships in which the toilet area for the regular sailors was placed at the head or bow of the ship.

### DESIGN

In sailing ships the toilet was placed in the bow for two reasons. First, since most vessels of the era could not sail directly into the wind, the winds came mostly across the rear of the ship placing the head essentially downwind. Secondly, if placed somewhat above the water line, vents or slots cut near the floor level would allow normal wave action to wash out the facility. Only the captain had his private toilet near his quarters at the stern of the ship in the quarter gallery.

In many modern boats, the heads look similar to a seated, flush toilets but use a system of valves and pumps that brings sea water into the toilet and pumps the waste out through the hull in place of the more normal cistern and plumbing trap to a drain.

In small boats the pump is often hand operated. The cleaning mechanism is easily blocked if too much toilet paper or other fibrous material is put down the pan.

Submarine heads face the problem that at greater depths higher water pressure makes it harder to pump the waste out through the hull. As a result early systems could be complicated, with the head fitted to the US S class being described as almost taking an engineer to operate.

Making a mistake resulted in the waste being expelled back into the body of the submarine. The toilet on the World War 1 British E class was considered so poor by the captain of HMS E35 that he preferred the crew to wait to relieve themselves until the submarine surfaced at night. As a result many submarines used the heads as an extra storage space for provisions.

## GUNROOM

A gunroom is the junior officers' mess on a naval vessel. It was occupied by the officers below the rank of lieutenant, but who are not warrant officers of the class of the boatswain, gunner or carpenter. In the wooden sailing ships it was on the lower deck, and was originally the quarters of the gunner, but in its form as a mess, guns were not normally found in it. The senior officers' equivalent is the wardroom.

In large ships of war, the gunroom was a compartment originally occupied by the gunner and his mates, but now fitted up for the accommodation of the junior officers; in smaller vessels, that used as a mess-room by the lieutenants.

'The Gunroom', named after this cabin, is also a longstanding online discussion group dedicated to the works of Patrick O'Brian, who wrote the Aubrey–Maturin series of novels about naval vessels and the characters found aboard them.

In an English country house, the Gun Room is a secure walk-in vault in which sporting rifles, shotguns, ammunition and other shooting accessories are kept. They are locked away partly for security, partly as some makes such as Holland and Holland or Purdey are highly valuable (costing as much as £60,000 for shotguns and £100,000 for rifles and with a 2-3 year waiting list from order to delivery).

## GUN DECK

The term gun deck originally referred to a deck aboard a ship that was primarily used for the mounting of cannon to be fired in broadsides. However, on many smaller vessels such as frigates and unrated vessels, the upper deck, forecastle and quarterdeck bore all of the cannons but were not referred to as the gun deck. The completely covered level under the upper deck was, however, still called the gun deck although it had no guns at all.

On board marine seismic survey vessels, the lowest deck on the ship is normally referred to as the gun deck. This deck carries the seismic source arrays, consisting of air guns arranged in clusters.

### SLANG

The term "gun deck" is also navy slang for fabricating or falsifying something. Some say the origin of the term dates to the practice of painting images of cannon ports on the side of a ship to make it appear to have more guns than it actually did, and thereby convincing any adversaries that they were outgunned, forgoing engagement.

A more plausible explanation relates to midshipmen retiring to the gun deck to complete their celestial navigation assignments of computing the ship's position three times daily following morning star sights, noon sun line, and evening star sights. While some midshipmen might be conscientious about computing positions from new observations, others were reputed to extrapolate and back calculate observation data from dead reckoning courses and speeds since earlier observations; and the actual computations performed on the gun deck were suspect.

This term is presently used to indicate the falsification of documentation in order to avoid actually doing the work or make present conditions seem otherwise acceptable.

## GALLEY (KITCHEN)

The galley is the compartment of a ship, train or aircraft where food is cooked and prepared. It can also refer to a land based kitchen on a naval base or a particular formed household kitchen.

### SHIP'S KITCHEN

A galley is the kitchen aboard a vessel, usually laid out in an efficient typical style with longitudinal units and overhead cabinets. This makes the best use of the usually limited space aboard ships. It also caters for the rolling and heaving nature of ships, making them more resistant to the effects of the movement of the ship.

For this reason galley stoves are often gimballed, so that the liquid in pans does not spill out. They are also commonly equipped with bars, preventing the cook from falling against the hot stove.

A small kitchen on deck was called a caboose or camboose, originating from the Dutch: *kombuis* which is still in use today. In English it is a defunct term used only for a cooking area that is abovedecks.

## HOUSEHOLD KITCHEN

The term *Galley kitchen* is also used to refer to the design of household kitchen wherein the units are fitted into a continuous array with no kitchen table, allowing maximum use of a restricted space, and work with the minimum of required movement between units.

Such kitchens increase storage space by working vertically, with hanging pots, dish racks, and ceiling-hung cabinets common. Strictly, the term refers to a kitchen with the units in two facing lines, but is often used to refer to U-shaped kitchens as well.

The first mass-produced galley kitchen design was known as the Frankfurt kitchen, designed by Margarete Schütte-Lihotzky, working under the direction of Ernst May in 1926 for a Frankfurt housing estate. 10,000 units were installed in Frankfurt, and it was the most successful and influential kitchen of the period.

## AVIATION KITCHEN

The Douglas Aircraft DC-3 was the first airplane with a planned galley for food service. Galleys on commercial airlines typically include not only facilities to serve and store food and beverages, but also contain flight attendant jumpseats, emergency equipment storage, as well as anything else flight attendants may need during the flight.

Aircraft in operation today mainly use the familiar trolley system. This system was introduced in the late 1960s at the same time as a new generation of large "widebody" aircraft were entering into service with the airlines. The significantly larger number of passengers on these aircraft meant that meals could no longer be efficiently delivered by hand, as they had been up until that time.

There are two main sizes of trolley in use with the airlines around the world, called the "ATLAS" and "KSSU" sizes. ATLAS is by far the most common size with approximately 80% of airlines using it. Remaining users of KSSU include KLM and Cathay Pacific.

Airbus has developed a new galley concept called SPICE, presented 2010, which they publicise as a potential new worldwide standard with significant advantages over the current 40-year-old trolley technology.

## WERNER SELL

The first airplane kitchen was invented (patent) by Werner Sell (Georg Robert Werner Sell) of Germany in 1930; the first fitted kitchens were delivered 1954, by 1955 Lufthansa had all their commercial planes fitted with a Sell galley. In 1955 Sell also began fitting train coaches with kitchens, from 1960 on with the newly developed convection oven. The Frankfurt kitchen needed an exact space, whereas the Sell kitchen could be fitted anywhere.

## FLYING BRIDGE

A flying bridge is an open area on top of a surface ship which provides unobstructed views of the fore, aft, and the sides of a vessel, and which serves as an operating station for the ship's officers, such as the captain or officer of the watch.

### SURFACE VESSELS

Prior to World War II, virtually every sailing ship, steamship, monitor, paddle steamer, or large pleasure ship had a flying bridge above the pilot house or main bridge. Flying bridges were generally not enclosed at all (although sometimes they were partially enclosed), and often had little equipment—usually just a speaking tube or telephone to allow communication with the helmsman or wheelman on the main bridge. On military warships after 1914, the flying bridge was usually the station for the air defense officer and the gunnery officer.

The amount of equipment on a flying bridge varies widely with the need of the captain. During World War II, for example, American submarine chaser surface ships had a well-outfitted flying bridge which usually contained a pelorus, signal lamps, telescope, and voice tube to permit the captain to command the ship. U.S. Navy attack transport ships could be outfitted with either 20mm or 50mm machine guns on their flying bridges. Flying bridges were almost always the highest bridge on the ship. They usually were above the flag bridge (also known as the "admiral's bridge") and the main bridge.

Since the 1980s, the pilot house on smaller craft usually consists of a helm chair directly over the center line of the vessel; electronics which monitor all systems about the vessel; engine monitoring systems and controls; and other electronics such as radar, radio, depth sounders, and so on. On the smallest surface vessels, such as a sport fishing boat, the flying bridge may have controls permitting the ship to be piloted from the flying bridge, but will lack the full range of controls of the pilot house. On larger small vessels, the flying bridge may actually be enclosed, in which case it is more properly called an "upper pilot house" or "upper bridge".

Some flying bridges have "bridge wings", open areas which thrust outward from the flying bridge over the sides of the vessel by approximately 10 to 15 feet (3.0 to 4.6 m) to allow an officer to see round the cabin or smokestack behind the flying bridge.

### SUBMARINES

On a submarine, the flying bridge is usually called "the bridge". It is the highest point on the conning tower, to provide for better visual navigation when on the surface. They became standard on United States Navy submarines after 1917, and greatly improved the function of the ships while at the surface.

## FIRE ROOM

On a ship, the fire room, or FR or boiler room or stokehold, referred to the space of a vessel where water was brought to a boil. The steam was then transmitted to a separate engine room, located immediately aft, where it was utilized to power the vessel.

To increase the safety and damage survivability of a vessel, the machinery necessary for operations may be segregated into various spaces, the fire room was one of these spaces, and was among the largest physical compartment of the machinery space. On some ships, the space comprised more than one fire room, such as forward and aft, or port or starboard fire rooms, or may be simply numbered. Each room was connected to a stack ventilating smoke. By their nature, fire rooms were less complex than their allied engine room and were normally supervised by less senior personnel.

On a large percentage of vessels, ships and boats, the fire room was located near the bottom, and at the rear, or aft, end of the vessel, and usually comprised few compartments. This design maximized the cargo carrying capacity of the vessel. The fire room on some ships was situated amid-ships, especially on vessels built from the 1880s to the 1990s.

### EQUIPMENT

Vessels typically contained several engines for different purposes. Main, or propulsion engines are used to turn the ship's propeller and move the ship through the water. Their allied fire room typically burned heavy fuel oil, replacing the earlier use of coal. There was a mechanism for removing ash from combustion that did not rise out of the stack.

On a steamship, power for both electricity and propulsion is provided by one or more large boilers giving rise to the alternate name boiler room. The latter name was preferred in the British Navy, among others. High pressure steam from the boiler is piped to the engine room to drive reciprocating engines or turbines for propulsion, and turbo generators for electricity.

Naval ships typically were able to generate a large volume of smoke by changing the fuel mix. Prior to the heavy use of radar, a smoke screen could be used to mask the movement of ships.

### DAMAGE CONTROL

Damage control was enhanced by the separation of the fire and engine rooms. In the event of damage to its associated engine room, steam could be transmitted to another engine room. In turn, an engine room could still operate though its associated fire room had become inoperative.

Two engineering advances resulted in the disappearance of the fire room in the early 1990s. The first was the movement by naval shipbuilding to nuclear-powered vessels. If a room containing nuclear material was subjected to damage, it was assumed that the event would likely result in abandonment of the ship regardless of the separation of rooms.

The second was the adoption of gas turbines in place of oil-fired boilers for all other navy ships. These powered engines directly and needed no boilers.

## SAFETY

### Fire precautions

Fire rooms were hot, sometimes dirty, and potentially dangerous. The presence of flammable fuel meant that a fire hazard existed in the fire room, which was monitored continuously by the ship's engineering staff and various monitoring systems.

### Ventilation

Fire rooms employed some means of providing air for the operation of the flame to ignite the oil and associated ventilation. Only spot ventilation was practical to keep personnel cool. This would require an unrestricted hull opening of the same size as the intake area of the engine itself assuming the hull opening is in the fire room itself.

Forced draft fire rooms were used until World War II. These required that personnel enter through an air lock to maintain the pressure. These were abandoned when the forced draft occasionally failed and blowback occurred killing fire room personnel.

Commonly, screens were placed over openings reducing airflow by approximately 50% so the opening area was increased appropriately. The requirement for general ventilation and the requirement for sufficient combustion air are quite different. A typical arrangement might be to make the opening large enough to provide intake air plus 1,000 cubic feet (28 m) per Minute (CFM) for additional ventilation. Engines pull sufficient air into the fire room for their own operation. However, additional airflow for ventilation usually requires intake and exhaust blowers.

## STAFFING REQUIREMENTS

When fired up, there were personnel assignments specified underway, as well as in port. For example, for an *Iowa*-class battleship, in normal steaming four boilers were operated. This was sufficient to power the ships at speeds up to 27 knots (50 km/h). For higher speeds, all eight boilers were lit. Each operating boiler required a minimum of four trained operators on watch: a boiler supervisor (BTOW), a superheater burnerman and saturated burnerman to control the steam temperature and pressure and a checkman, who monitored and controlled the water level in the steam drum. In addition, there was a fireroom messenger and a lower level pumpman on duty whenever the fireroom was steaming.

## ENGINE ROOM

On a ship, the engine room, or ER, is the propulsion machinery spaces of the vessel. To increase the safety and damage survivability of a vessel, the machinery necessary for operations may be segregated into various spaces. The engine room is one of these spaces, and is generally the largest physical compartment of the machinery space.

The engine room houses the vessel's prime mover, usually some variations of a heat engine - diesel engine, gas or steam turbine. On some ships, the machinery space may comprise more than one engine room, such as forward and aft, or port or starboard engine rooms, or may be simply numbered.

On a large percentage of vessels, ships and boats, the engine room is located near the bottom, and at the rear, or aft, end of the vessel, and usually comprises few compartments.

This design maximizes the cargo carrying capacity of the vessel and situates the prime mover close to the propeller, minimizing equipment cost and problems posed from long shaft lines. The engine room on some ships may be situated mid-ship, especially on vessels built from 1900 to the 1960s. With the increased use of diesel electric propulsion packages, the engine room(s) may be located well forward, low or high on the vessel, depending on the vessel use.

### EQUIPMENT

#### Engines

The engine room of a motor vessel typically contains several engines for different purposes. Main, or propulsion engines are used to turn the ship's propeller and move the ship through the water. They typically burn diesel oil or heavy fuel oil, and may be able to switch between the two. There are many propulsion arrangements for motor vessels, some including multiple engines, propellers, and gearboxes.

Large engines drive electrical generators that provide power for the ship's electrical systems. Large ships typically have three or more synchronized generators to ensure smooth operation.

The combined output of a ship's generators is well above the actual power requirement to accommodate maintenance or the loss of one generator. On a steamship, power for both electricity and propulsion is provided by one or more large boilers giving rise to the alternate name boiler room. High pressure steam from the boiler is used drive engines: reciprocating engines or turbines for propulsion, and turbo generators for electricity.

Besides propulsion and auxiliary engines, a typical engine room contains many smaller engines, including generators, air compressors, feed pumps, and fuel pumps. Today, these machines are usually powered by small diesel engines or electric motors, but may also use low-pressure steam.

### Engine cooling

The engine(s) get required cooling from liquid-to-liquid heat exchangers connected to fresh seawater or divertible to recirculate through tanks of seawater in the engine room. Both supplies draw heat from the engines via the coolant and oil lines.

Heat exchangers are plumbed in so that oil is represented by a yellow mark on the flange of the pipes, and relies on paper type gaskets to seal the mating faces of the pipes. Sea water or brine, is represented by a green mark on the flanges and internal coolant is represented by blue marks on the flanges.

### Thrusters

In addition to this array of equipment is the ships thruster system, typically operated by electric motors controlled from the bridge. These thrusters are laterally mounted propellers that can suck or blow water from port to starboard (i.e. left to right) or vice versa. They are normally used only in maneuvering, e.g. docking operations, and are often banned in tight confines, e.g. drydocks.

Thrusters, like main propellers, are reversible by hydraulic operation. Small embedded hydraulic motors rotate the blades up to 180 degrees to reverse the direction of the thrust.

## SAFETY

### Fire precautions

Engine rooms are hot, noisy, sometimes dirty, and potentially dangerous. The presence of flammable fuel, high voltage (HV) electrical equipment and internal combustion engines (ICE) means that a serious fire hazard exists in the engine room, which is monitored continuously by the ship's engineering staff and various monitoring systems.

### Ventilation

If equipped with internal combustion or turbine engines, engine rooms employ some means of providing air for the operation of the engines and associated ventilation. If individuals are normally present in these rooms, additional ventilation should be available to keep engine room temperatures to acceptable limits. If personnel are not normally in the engine space, as in many pleasure boats, the ventilation need only be sufficient to supply the engines with intake air.

This would require an unrestricted hull opening of the same size as the intake area of the engine itself, assuming the hull opening is in the engine room itself. Commonly, screens are placed over such openings and if this is done, airflow is reduced by approximately 50%, so the opening area is increased appropriately.

The requirement for general ventilation and the requirement for sufficient combustion air are quite different. A typical arrangement might be to make the opening large enough to provide intake air plus 1000 Cubic Feet per Minute (CFM) for additional ventilation. Engines pull sufficient air into the engine room for their own operation. However, additional airflow for ventilation usually requires intake and exhaust blowers.

### HISTORY

Engine rooms were separated from its associated fire room on fighting ships from the 1880s through the 1960s. If either experienced damage putting it out of action, the associated engine room could get steam from another fire room.

## DECK (SHIP)

A deck is a permanent covering over a compartment or a hull of a ship. On a boat or ship, the primary or upper deck is the horizontal structure which forms the 'roof' for the hull, which both strengthens the hull and serves as the primary working surface.

Vessels often have more than one level both within the hull and in the superstructure above the primary deck which are similar to the floors of a multi-storey building, and which are also referred to as decks, as are specific compartments and decks built over specific areas of the superstructure.

### STRUCTURE

The purpose of the upper or primary deck is structural, and only secondarily to provide weather-tightness, and to support people and equipment. The deck serves as the lid to the complex box girder which is the hull. It resists tension, compression, and racking forces.

The deck's scantling is usually the same as the topsides, or might be heavier if the deck is expected to carry heavier loads (for example a container ship). The deck will be reinforced around deck fittings such as the capstan, cleats, or bollards.

On ships with more than one level, deck refers to the level itself. The actual floor surface is called the sole, the term deck refers to a structural member tying the ships frames or ribs together over the keel. In modern ships, the interior decks are usually numbered from the primary deck, which is #1, downward and upward.

So the first deck below the primary deck will be #2, and the first above the primary deck will be #A2 or #S2 (for "Above" or "Superstructure"). However, ships may also call decks by common names, or (especially on cruise ships) may invent fanciful and romantic names for a specific deck or area of that specific ship, such as the *Lido deck* of the Princess Cruises' *Love Boat*.

Equipment mounted on deck, such as the ship's wheel, binnacle, fife rails, and so forth, may be collectively referred to as deck furniture. Weather decks in western designs evolved from having structures fore and aft (forward or front and aft or The Rear of the ship mostly clear, then in the 19th century pilothouses/wheelhouses and deckhouses began to appear, eventually developing into the superstructure of modern ships. Eastern designs developed earlier, with efficient middle decks and minimalist fore and aft cabin structures across a range of designs.

## COMMON NAMES FOR DECKS

In vessels having more than one deck there are various naming conventions, numerically, alphabetically, etc. However, there are also various common historical names and types of decks:

- 01 level is the term used in naval services to refer to the deck above the main deck. The next higher decks are referred to as the 02 level, the 03 level, and so on. Although these are formally called decks, they are usually referred to as levels, because they are usually incomplete decks that do not extend all the way from the stem to the stern or across the ship.
- Berth deck: (*Naval*) A deck next below the gun deck, where the hammocks of the crew are slung.
- Boat deck: Especially on ships with sponsons, the deck area where lifeboats or the ship's gig are stored.
- Boiler deck: (river steamers) The deck on which the boilers are placed.
- Bridge deck: (a) The deck area including the helm and navigation station, and where the Officer of the Deck/Watch will be found, also known as the *conn* (b) An athwartships structure at the forward end of the cockpit with a deck, often somewhat lower than the primary deck, to prevent a pooping wave from entering through the companionway. May also refer to the deck of a bridge.
- Flight deck: (*Naval*) A deck from which aircraft take off or land.
- Flush deck: Any continuous unbroken deck from stem to stern.
- Forecastle deck: The foremost part of the upper deck under which the sailors have their berths, extending from the foremast to the bow.

- Gun deck: (*Naval*) a deck below the spar deck where the ships' cannon were carried. The term originally referred to a deck aboard a ship that was primarily used for the mounting of cannon to be fired in broadsides. However, on many smaller vessels such as frigates and unrated vessels the upper deck, forecastle and quarterdeck bore all of the cannons but were not referred to as the gun deck. The completely covered level under the upper deck was, however, still called the gun deck although it had no guns at all.
- Half-deck: That portion of the deck next below the spar deck which is between the mainmast and the cabin.
- Helicopter deck (helo deck): Usually located near the stern and always kept clear of obstacles hazardous to a helicopter landing.
- Hurricane deck: (River Steamers, etc.), the upper deck, usually a light deck, erected above the frame of the hull (deriving its name from the wind that always seemed to blow on the deck).
- Lido deck: Open area, typically at or near the stern of a passenger ship, housing the main outdoor swimming pool and sunbathing area.
- Lower deck: (a) the deck immediately over the hold, orig. only of a ship with two decks. (b) synonym for berth deck.
- Main deck: The principal deck of a vessel; in some ships the highest deck of the hull, usually but not always the weather deck; in sailing warships often a deck under the upper deck.
- Middle or Waist deck The upper deck amidships, the working area of the deck.
- Orlop deck: The deck or part of a deck where the cables are stowed, usually below the waterline. It is the lowest deck in a ship.
- Poop deck: The deck forming the roof of a poop or poop cabin, built on the upper deck and extending from the mizzenmast aft.
- Promenade deck: A "wrap-around porch" found on passenger ships and riverboats encircling the superstructure. This can have open railings or be enclosed in glass, or a combination. Often the entire level where this is located is referred to as the Promenade Deck.
- Quarterdeck: (a) The part of the upper deck abaft the mainmast, including the poop deck when there is one. Usually reserved for ship's officers, guests, and passengers. (b) (*Naval*) The area to which a gangway for officers and diplomatic guests to board the vessel leads. Also any entry point for personnel.
- Side-deck: The upper deck outboard of any structures such as a coachroof or doghouse, also called a breezeway
- Spar deck: (a) Same as the upper deck. (b) Sometimes a light deck fitted over the upper deck.
- Sweep deck: (*Naval*) The aftmost deck on a minesweeper, set close to the waterline for ease in launch and recovery of equipment.

- Topgallant forecastle deck: Any raised deck occurring above the forecastle deck.
- Tween deck: the storage space between the hold and the main deck, often retractable.
- Upper deck: The highest deck of the hull, extending from stem to stern.
- Weather deck: (a) Any deck exposed to the outside. (b) The windward side decks.
- Well deck: (a) An exposed deck (weather deck) lower than decks fore and aft. In particular it is one enclosed by bulwarks limiting flow of water and thus drainage. (b) (*Modern naval*) A hangar-like deck located at the waterline in the stern of some amphibious assault ships, also known as a well dock. By taking on water the ship can lower the stern flooding the well deck and allowing boats and landing craft to dock within the ship.

## CONSTRUCTION

### Methods in wood

A traditional wood deck would consist of planks laid fore and aft over beams and along carlins, the seams of which are caulked and paid with tar. A yacht or other fancy boat might then have the deck canvased, with the fabric laid down in a thick layer of paint or sealant, and additional coats painted over. The wash or apron boards form the joint between the deck planking and that of the topsides, and are caulked similarly.

Modern "constructed decks" are used primarily on fiberglass, composite, and cold-molded hulls. The under structure of beams and carlins is the same as above. The decking itself is usually multiple layers of marine-grade plywood, covered over with layers of fibreglass in a plastic resin such as epoxy or polyester overlapped onto the topsides of the hull.

### Methods in metal

Generally speaking, the method outlined for "constructed decks" is most similar to metal decks. The deck plating is laid over metal beams and carlins and tacked temporarily in place. The difficulty in metal construction is avoiding distortion of the plate while welding due to the high heat involved in the process. Welds are usually double pass, meaning each seam is welded twice, a time consuming process which may take longer than building the wood deck. But welds result in a waterproof deck which is strong and easily repairable. The deck structure is welded to the hull, making it structurally a single unit. Because a metal deck, painted to reduce corrosion, can be quite slippery, pick up heat from the sun, and be quite loud to work on, a layer of wood decking or thick non-skid paint is often applied to its surface.

## Methods in fiberglass

The process for building a deck in fiberglass is the same as for building a hull: a female mould is built, a layer of gel coat is sprayed in, then layers of fiberglass in resin are built up to the required deck thickness (if the deck has a core, the outer skin layers of fiberglass and resin are laid, then the core material, and finally the inner skin layers.) The deck is removed from the mould and usually mechanically fastened to the hull.

Fiberglass decks are quite slick with their mirror-smooth surfaces, so a non-skid texture is often moulded into their surface, or non-skid pads glued down in working areas.

## Rules of thumb to determine the deck scantlings

The thickness of the decking affects how strong the hull is, and is directly related to how thick the skin of the hull itself is, which is of course related to how large the vessel is, the kind of work it is expected to do, and the kind of weather it may reasonably be expected to endure. While a Naval Engineer or Architect may have precise methods of determining what the scantlings should be, traditional builders used previous experiences and simpler rules-of-thumb to determine how thick the deck should be built. The numbers derived by these formulae gives a rough number for determining the average thickness of materials based on some crude hull measurements. Below the waterline the thickness should be approximately 115% of the result, while upper topsides and decks might be reduced to 85% of the result.

- In wood – For plank thickness in inches, LOA (Length OverAll) and Beam are measured in feet. For plank thickness in mm, LOA and Beam are measured in meters.
  - Plank thickness in inches = $\dfrac{\sqrt{LOA} + Beam}{16}$
  - Plank thickness in mm = $[\sqrt{LOA{\cdot}3.28} + (Beam{\cdot}3.28){\cdot}1.58]$
- In fiberglass – For skin thickness in inches, LWL (Length WaterLine) is in feet. For skin thickness in mm, LWL is in meters.
  - Skin thickness (inches) = $0.07 + \dfrac{LWL}{150}$
  - Skin thickness (mm) = $1.8 + \dfrac{LWL}{1.8}$
- In fiberglass sandwich – First determine the skin thickness as single skin, then multiply by modifiers for inner skin, outer skin, and core thicknesses. Cored decks might be modified even thicker, 2.6–2.7, to increase stiffness.
  - Inner skin modifier = 0.3
  - Outer skin modifier = 0.4
  - Core modifier = 2.2

## CABOOSE (SHIP'S GALLEY)

Caboose (also *camboose, coboose, cubboos* derived from the Dutch *kombuis*) is a term used for a small ship's kitchen, or galley on deck. At one time a *caboose* related to a smaller kitchen on a merchantman ship, while on a larger warship it was called a *galley*. William Falconer's 1780 *A Universal Dictionary of the Marine* describes a caboose thus: "a sort of box or house to cover the chimney of some merchant-ships.

It somewhat resembles a sentry-box, and generally stands against the barricade on the fore part of the quarter-deck". Sometimes the caboose was portable. Previous to the introduction of the caboose the furnaces for cooking were, in three-deckers, placed on the middle deck, in two-decked ships in the forecastle. In Canada a caboose was formerly used to describe a galley on a timber raft. The term was sometimes also applied to the cast-iron stove used for cooking on deck or in galleys during the early 19th century, as well as an outdoor oven or fireplace.

## CABIN (SHIP)

A cabin or berthing is an enclosed space generally on a ship or an aircraft. A cabin which protrudes above the level of a ship's deck may be referred to as a "deckhouse."

### SAILING SHIPS

In sailing ships, the officers and paying passengers would have an individual or shared cabin. The captain or commanding officer would occupy the "great cabin" that normally spanned width of the stern with large windows. On a warship it would be separated from the rest of the ship, and further subdivided into daytime and night-time cabins with movable panels that could be removed down in time of battle to leave the deck clear the whole length of the ship.

### MODERN WARSHIPS

In most modern warships the commanding officer has a main cabin, the in-port cabin, often adjacent to the ship's central control room (operations room), and a sea cabin adjacent to the bridge. Thus, when likely to be called from sleep or attending to administration, the CO can be at the Bridge or Ops room instantly.

The sea cabin is sparsely equipped, containing just a bunk, a desk, and basic toilet facilities. The in-port cabin is more lavishly furnished, with separate bedroom and combination sitting room/office, and more elaborate toiletry facilities.

For ships intended to act as flagships, like the aircraft carrier the USS Lexington (CV-16), the admiral also has a sea cabin (adjacent to the captain's sea cabin) and an in-port cabin, in addition to the captain. Admiral Fletcher's sea cabin in the USS Yorktown (CV-5) in the Second World War had a bed, an easy chair, a table, and a shower.

Officers will normally have their own cabins, which doubles as their office. Some senior petty officers may have a cabin for similar reasons.

### PASSENGER SHIPS

In ships carrying passengers, they are normally accommodated in cabins, taking the terminology familiar to seafarers. First class cabins were traditionally referred to as staterooms, and today many cruise lines now prefer to refer to passenger cabins as staterooms or suites.

### CABIN CRAWL

In cruise ship terms, a cabin crawl is an event where passengers tour the cabins of fellow passengers.

A cruise ship may also offer a cabin crawl of cabins or suites which did not sell for a particular sailing. The purpose of a cabin crawl is to give passengers an idea of the space and layout of various cabin options for their next cruise. Cabin crawls are normally organized prior to a cruise through cruise fan websites.

### SPACECRAFT

In ships going into space the cabins are required to fully supply food and Oxygen for their crew. On missions lasting a year or longer the cabins have to be self-sustaining, i.e. replenish their own water and oxygen. The space cabin for any long-range manned mission is expected to be reasonably spacious, with approximately 28 cubic metres allotted to each occupant. In addition cabins have life support systems that should have the capability to meet a variety of off-nominal conditions, including cabin fires, depressurization, and component shutdown or failure.

Frequently these conditions occur so quickly that automatic control systems offer the only possibility for recovery. In the late 1960s several experimental ground facilities were developed to evaluate regenerative life support systems for manned space flight.

## BRIDGE (NAUTICAL)

The bridge of a ship is the room or platform from which the ship can be commanded. When a ship is underway the bridge is manned by an OOW (officer of the watch) aided usually by an AB (able seaman) acting as lookout. During critical maneuvers the captain will be on the bridge supported, perhaps, by an OOW as an extra set of hands, an AB on the wheel and sometimes a pilot if required.

Wheelhouses are the small enclosed parts of a bridge which historically held the ship's steering wheel. Today, ship bridges do not have a separate wheelhouse. In some cases, the term wheelhouse is used interchangeably to refer to the bridge.

## EVOLUTION

Traditionally, sailing ships were commanded from the quarter deck, aft of the mainmast. With the arrival of paddle steamers, engineers required a platform from which they could inspect the paddle wheels and where the captain's view would not be obstructed by the paddle houses. A raised walkway, literally a bridge, connecting the paddle houses was therefore provided. When the screw propeller superseded the paddle wheel, the bridge was retained.

Traditionally, commands would be passed from the senior officer on the bridge to stations dispersed throughout the ship, where physical control of the ship was exercised, as technology did not exist for the remote control of steering or machinery.

Helm orders would be passed to an enclosed wheel house, where the coxswain or helmsman operated the ship's wheel. Engine commands would be relayed to the engineer in the engine room by an engine order telegraph, which displayed the captain's orders on a dial. The engineer would ensure that the correct combination of steam pressure and engine revolutions were applied. The bridge was often open to the elements, therefore a weatherproof pilot house could be provided, from which a pilot, who was traditionally the ship's navigating officer, could issue commands from shelter.

Iron, and later steel, ships also required a compass platform. This was usually a tower, where a magnetic compass could be sited far away as possible from the ferrous interference of the hulk of the ship.

Depending upon the design and layout of a ship, all of these terms can be variously interchangeable. Many ships still have a flying bridge, a platform atop the pilot house, open to weather, containing a binnacle and voice tubes to allow the conning officer to direct the ship from a higher position during fair weather conditions.

Larger ships, particularly warships, often had a number of different bridges. A navigation bridge would be used for the actual conning of the ship. A separate admiral's bridge could be provided in flagships, where the admiral could exercise strategic control over his fleet without interfering with the Captain's tactical command of the vessel. In older warships, a heavily armored conning tower was often provided, where the vital command staff could be located under protection to ensure that the ship could be commanded under fire.

Modern advances in remote control equipment have seen progressive transfer of the actual control of the ship to the bridge. The wheel and throttles can be operated directly from the bridge, controlling often-unmanned machinery spaces. Aboard modern warships it is common that actual commanding is carried out on the bridge while weapon systems are controlled from a room usually deep inside the vessel. This also enables a redundant setup in which the vessel is still able to fight in case one of the command instances has been hit or destroyed.

## BILGE

The bilge is the lowest compartment on a ship, below the waterline, where the two sides meet at the keel. The first known use of the word is from 1513.Bilge water

The word is sometimes also used to describe the water that collects in this area. Water that does not drain off the side of the deck drains down through the ship into the bilge. This water may be from rough seas, rain, leaks in the hull or stuffing box, or other interior spillage. The collected water must be pumped out to prevent the bilge from becoming too full and threatening to sink the ship.

Bilge water can be found aboard almost every vessel. Depending on the ship's design and function, bilge water may contain water, oil, urine, detergents, solvents, chemicals, pitch, particles, and other materials.

By housing water in a compartment, the bilge keeps these liquids below decks, making it safer for the crew to operate the vessel and for people to move around in heavy weather.

### BILGE MAINTENANCE

Methods of removing water from bilges have included buckets and pumps. Modern vessels usually use electric bilge pumps usually controlled by automated bilge switches. Bilge coatings are applied to protect the bilge surfaces.

The water that collects is often noxious, and "bilge water" or just "bilge" has thus become a derogatory colloquial term used to refer to something bad, fouled, or otherwise offensive.

Bilges may contain partitions to damp the rush of water from side to side and fore and aft to avoid destabilizing the ship due to the free surface effect. Partitions may contain limber holes to allow water to flow at a controlled rate into lower compartments.

Cleaning the bilge and bilge water is also possible using "passive" methods such as bioremediation, which uses bacteria to break down the hydrocarbons in the bilge water into harmless byproducts. Of the two general schools of thought on bioremediation, the one that uses bacteria local to the bilge is regarded as being more "green" because it does not introduce foreign bacteria to the waters that the vessel sits in or travels through.

The term "bilged" refers to the deliberate or accidental flooding of the bilge which incapacitates the ship, slows its speed, and/or to makes it difficult or impossible to handle. This process may be carried out intentionally by enemy action, or by the crew of the ship itself in order to save it from falling into enemy hands.

Taking this action may allow the ship to be recovered later by being pumped out and re-floated during a high tide. When done on purpose it is normally referred to as scuttling. When done accidentally by grounding, it usually resulted in the ship being entirely lost.

# 5

# Ship Disposal

A number of different methods exist for disposing of a ship after it has reached the end of its effective or economic service life with an organisation.

## OPTIONS

Although many options are technically possible, some, such as deep water sinking, are not used for non-military vessels, or have come under increased scrutiny. Options currently available include:

- Hulking
- Ship breaking (involving recycling)
- Use as an artificial reef
- Donation
- Sale for re-use
- Floating (or drydock) storage
- Deep water sinking

## DESCRIPTION OF OPTIONS

- Hulking was a traditional method of converting a hull to another purpose after its usefulness as a ship had ended. The ship is stripped of its motive equipment (sails and rigging or motors) and is used for a variety of purposes. This practice is still in use to a limited extent.
- Ship breaking is the most common and most environmentally accepted method of ship disposal. According to various organisations, only facilities approved by the Basel Action Network's "Green Ship Recycling" program are environmentally sound options.
- Artificial reefing is the sinking of ships offshore to form reefs. Before sinking, the vessel needs to have all toxic components and electrical devices removed. Depending on the ship design this may be done relatively easily, although some designs present certain difficulties.

Also the reef location needs to be determined. For these, approval must be obtained from the state authorities. According to the Rand Corporation, artificial reefing to form wreck diving sites is however still a profitable option, as the reefs are anticipated to be visited by tourists, thus generating an income.

- Donation, or sale for re-use, allows for the refurbishment and reuse of a vessel by a secondary party. Such uses can include further life as a ship, houseboat, office, hotel, museum, or conference center.
- Deep water sinking or SINKEX is the sinking of ships during target practice. Before the ships are sunk they are cleaned of all toxic components and devices, so that no environmental pollution should occur.
- Floating or drydock storage involves storing the boat on or off water for a period of time. Storage on water is an expensive method as after 30 years the ships need extensive repairs due to corrosion by the seawater. Storage in drydocks is also expensive. Regular storage on shore, outside of a dock, is also a possibility.

## ILLEGAL SHIP DISPOSAL

Some shipping companies have in the past attempted to dispose of their ships illegally, often due to the high costs of sanitizing these ships in a legal way as a result of the presence of large quantities of toxic compounds and materials. Prominent examples of where environmental concerns have complicated the planned disposal of ships have been the Norwegian Cruise Lines vessels SS *Norway* and SS *Oceanic,* as well as the French aircraft carrier *Clemenceau*.

Although in the past there were few effective controls on ship disposal, and the risk of being caught was therefore limited, national authorities as well as environmental organisations are now more vigilant. Ships are sometimes tracked when they are nearing the end of their lifespan. Companies caught disposing of ships illegally face hefty fines.

## TARGET SHIP

A target ship is a vessel — typically an obsolete or captured warship — used as a seaborne target for naval gunnery practice or for weapons testing.

### RATIONALE

Sinking redundant warships is an effective way of testing new weapons and warships in as realistic a manner as possible. Whilst practice torpedoes are fired fairly frequently, they behave differently from warshots.

### PREPARATION

In order to meet environmental, health, and safety standards, ships have to be thoroughly cleaned so that all dangerous material and potential contaminants (such as asbestos, refrigerants etc.) are removed. In the event of the vessel becoming an artificial reef, escape exits also have to be created in the vessel, should divers encounter problems.

It is now also common practice to remove pennant numbers and sink the warships anonymously, as a mark of respect to those who sailed in them.

## NOTABLE EXAMPLES

### *Pacificateur*

In September 1819, the French engineer and Army artillery officer Henri-Joseph Paixhans wrote to the Ministry of the Navy to propose a fusing system to fire explosive shells at wooden warships, instead of the usual, solid round shots that were then in general naval use.

A commission studied the matter, and decided to build two Paixhans howitzers for trial purposes in 1822. In 1824, the 80-gun ship of the line *Pacificateur*, made redundant by the Bourbon Restoration, was condemned. She was a *Bucentaure* class two-decker of the same type as the French flagship of the Battle of Trafalgar.

The two prototypes were fired at her with devastating effect. This led to the adoption of the Paixhans gun in 1827. They were used to great effect at the Battle of San Juan de Ulua, to the interest of British and US observers, who announced the demise of wooden warships and the era of the ironclad.

### *Agamemmnon* and *Centurion*

The British pre-dreadnought battleship HMS *Agamemmnon* was converted to radio-control in 1920-21 and used for assessments of the damage that could be caused by aircraft and various calibres of guns. She was replaced in the role by the battleship *Centurion* in 1926.

### *Iowa*

After World War I ended, the US Navy and Army did live fire testing of attacking warships from the air. To get the testing as close to wartime conditions as possible, the USS *Iowa* was converted into a radio-controlled target ship, the first in United States naval history. A well known radio engineer, John Hays Hammond, Jr. developed the radio control gear for the Iowa. In 1923, the Iowa was sunk off the Pacific coast of Panama during fleet exercises, with members of Congress and the press attending, by the battleship *Mississippi*.

In the early 1930s the US Navy got serious about remote control ships and fitted the destroyer *Stoddert* with an improved radio controls developed by Lieutenant Commander Boyd R. Alexander, a radio design officer, and the *Naval Research Laboratory* in Bellevue D.C. for further testing and evaluation.

The evaluation proved so successful that the US Navy moved up their plans for radio controlled warships and in 1932 the obsolete battleship USS *Utah* and the destroyers *Boggs* and *Kilty* were converted to remote radio control.

### *James Longstreet*

A familiar sight for more than fifty years in Cape Cod Bay, Massachusetts was the SS *James Longstreet*. This World War II Liberty Ship was towed to a sandbar 3.5 miles (5.6 km) off shore in 1944 and was used for bombing practice through the Vietnam War.

### *Lützow*

The German Deutschland class cruiser, *Lützow* was refloated by the Soviet Navy after her scuttling in 1945 and used as a target in the Baltic Sea in 1949.

### Operation Crossroads

Operation Crossroads was a 1946 series of US nuclear tests at Bikini Atoll that used 95 target ships. Some were obsolete US ships, such as the USS *Nevada*, others were ships surrendered by the Axis powers at the end of World War II, such as the German heavy cruiser *Prinz Eugen* and the Japanese battleship *Nagato*.

### Operation Deadlight

Operation Deadlight was the destruction of 116 surrendered, German U-boats between November 1945 and February 1946 by the Royal Navy. The submarines were sunk in deep water to the north west of Ireland. Some were scuttled with explosive charges and others sank accidentally in bad weather while under tow. The remainder were sunk by ships and aircraft, using rockets, bombs, depth charges and gunfire.

### HMAS *Torrens*

The Royal Australian Navy (RAN) sunk HMAS *Torrens* on June 14, 1999 with a Mk48 wire guided torpedo fired from the *Collins* class submarine HMAS *Farncomb*. Torrens was the last of six Australian River class destroyers, the others (*Derwent*, *Parramatta*, *Stuart*, *Swan* and *Yarra*) having been disposed of previously. Before the sinking *Torrens* had been thoroughly cleaned of all fuels, oils and potentially environmentally harmful substances.

Her gun turret was donated to the South Western City of Albany. *Torrens* was then towed from Fleet Base West (HMAS *Stirling*) 90 kilometres (56 mi) out to sea, west of Perth. The submarine fired the torpedo at the stationary target from a submerged position over the horizon.

The sinking of *Torrens* was a display of firepower that provided some much needed positive publicity for the *Collins* class submarines, plagued by numerous technical problems and criticised over troubles with the combat system and noise reduction. Ric Shalders, commander of the Submarine Squadron said "the requirement of new submarine trials, the new need to test war-stock and the availability of the Torrens all came together to produce a very satisfactory result".

The US military term Sink Exercise (SINKEX) is used for the test of a weapons system usually involving a torpedo or missile attack of an unmanned target ship. The US Navy uses SINKEX to train its sailors on the usage of modern-day weapons.

This technique is used to dispose of decommissioned warships. The US Navy performs SINKEXs north of Kauai, Hawaii, in the Pacific Ocean off the coast of California, and near Puerto Rico.

## SHIP-SUBMARINE RECYCLING PROGRAM

The Ship/Submarine Recycling Program (SRP) is the process the United States Navy uses to dispose of decommissioned nuclear vessels. SRP takes place only at the Puget Sound Naval Shipyard (PSNS) in Bremerton, Washington, but the preparations can begin elsewhere. Before SRP can begin, the ship or submarine must have her nuclear fuel removed. Defueling usually coincides with decommissioning. Prior to that event, the vessel is referred to as "USS *Name*," but afterward the "USS" is dropped and it is referred to as "ex-*Name*."

Defueling of submarines is carried out at five ship repair facilities on the West Coast, and the hulls are then towed to PSNS. Reusable equipment is removed at the same time as the fuel. Spent nuclear fuel is shipped by rail to the Naval Reactor Facility in the Idaho National Laboratory (INL), located 42 miles (67 km) northwest of Idaho Falls, Idaho, where it is stored in special canisters.

At PSNS the SRP proper begins. A submarine is cut into three or four pieces: the aft section, the reactor compartment, the missile compartment if one exists, and the forward section. Missile compartments are dismantled according to the provisions of the Strategic Arms Reductions Treaty. Reactor compartments are sealed at both ends and shipped by barge and multiple-wheel high-capacity trailers to the Department of Energy's Hanford Nuclear Reservation in Washington state, where they are currently in open dry storage and slated to be eventually buried.

The burial trenches have been evaluated to be secure for at least 600 years before the first pinhole penetration of some lead containment areas of the reactor compartment packages occurs, and several thousand years before leakage becomes possible. In 1972, the London Dumping Convention restricted ocean disposal of radioactive waste. In 1993, ocean disposal of radioactive waste was completely banned. The US Navy began a study on scrapping nuclear submarines, two years later shallow land burial of reactor compartments was selected as the most suitable option. In 1990, USS Scamp (SSN 588) became the first US nuclear-powered submarine to have been successfully scrapped.

Prior to USS *Scamp*, in 1959, the US Navy removed a nuclear reactor from the submarine USS *Seawolf* (SSN 575) and replaced it with a new type. The removed reactor was scuttled in the Atlantic Ocean, 200 km east of Delaware, at a depth of 2,700 m. Until 1991, the forward and aft sections of the submarines were rejoined and placed in floating storage. Various proposals for disposal of those hulls were considered, including sinking them at sea, but none were economically practical.

All required removal of the numerous polychlorinated biphenyl products (PCBs) on board, which are considered hazardous materials by the Environmental Protection Agency and United States Coast Guard. Since then, and to help reduce costs, the remaining submarine sections are recycled, returning reusable materials to production. In the process of submarine recycling, all hazardous and toxic wastes are identified and removed, reusable equipment is removed and put into inventory.

Scrap metals and all other materials are sold to private companies or reused. The overall process is not profitable, but does provide some cost relief. Disposal of submarines by the SRP costs the US$25-50 million per submarine.

By the end of 2005, 195 nuclear submarines had been ordered or built in the US (including the NR-1 Deep Submergence Craft and *Virginia*, but none of the later *Virginias*). The last of the regular *Sturgeon* attack boats, *L. Mendel Rivers* was decommissioned in 2001, and *Parche*, a highly modified *Sturgeon*, was decommissioned in 2004.

The last of the initial "41 for Freedom" Fleet Ballistic Missile (FBM) submarines, *Kamehameha*, was decommissioned in 2002. Decommissioning of the *Los Angeles* boats began in 1995 with *Baton Rouge*. Additionally, a handful of nuclear-powered cruisers have entered the program, and their dismantling is ongoing.

The first aircraft carrier due for decommissioning that would enter the SRP is planned to be *Enterprise*, intended for withdrawal in 2013.

## SHIP GRAVEYARD

A ship graveyard or ship cemetery is a location where the hulls of scrapped ships are left to decay and disintegrate, or left in reserve. Such a practice is now less common due to waste regulations and so some dry docks where ships are dismantled (to recycle their metal and remove dangerous materials like asbestos) are also known as ship graveyards. By analogy, the phrase can also refer to a large number of shipwrecks which have accumulated in a single area but not been removed by human agency, instead being left to disintegrate naturally.

These can form in places where navigation is difficult or dangerous (such as the Goodwin Sands or Blackpool), where a large number of ships have been deliberately scuttled together (as with the Kaiserliche Marine at Scapa Flow), or where a large number of ships have been sunk in battle.

## SCUTTLING

Scuttling is the act of deliberately sinking a ship by allowing water to flow into the hull. This can be achieved in several ways—valves or hatches can be opened to the sea, or holes may be ripped into the hull with brute force or with explosives.

Scuttling may be performed to dispose of an abandoned, old, or captured vessel; to prevent the vessel from becoming a navigation hazard; as an act of self-destruction to prevent the ship from being captured by an enemy force; as a blockship to restrict navigation through a channel or within a harbor; or to provide an artificial reef for divers and marine life.

### NOTABLE HISTORICAL EXAMPLES

#### Hernán Cortés (1519)

The Spaniard conquistador Hernán Cortés, who led the first expedition that resulted in the fall of the Aztec empire, ordered his men to strip and scuttle his fleet in order to prevent the secretly planned return to Cuba by those loyal to Cuban Governor Diego Velázquez de Cuéllar.

Their success would have halted his inland march and conquest of the Aztec Empire.

#### HMS *Sapphire* (1696)

HMS *Sapphire* was a 32-gun, fifth-rate sailing frigate of the Royal Navy in Newfoundland to protect the English migratory fishery. The vessel was trapped in Bay Bulls harbour by four French naval vessels led by Jacques-François de Brouillan. In order to avoid capture, the English scuttled the vessel on 11 September 1696.

#### Russian Black Sea Fleet in Sevastopol (1854)

During the Crimean War, in anticipation of the Siege of Sevastopol, the Russians scuttled ships of the Black Sea Fleet to protect the harbour, to use their naval cannon as additional artillery, and to free up the ships' crews as marines.

Those ships that were deliberately sunk included *Grand Duke Constantine, City of Paris* (both with 120 guns), *Brave, Empress Maria, Chesme, Yagondeid* (84 guns), *Kavarna* (60 guns), *Konlephy* (54 guns), steam frigate *Vladimir*, steamboats

*Thunderer, Bessarabia, Danube, Odessa, Elbrose* and *Krein*. After the war, it became a technological challenge to recover the ships.

### USS *Merrimack*/CSS *Virginia* (1861)

In 1861, the Federal steam frigate *Merrimack* was set afire in Gosport Shipyard (now Norfolk Naval Shipyard) to keep her from falling into Confederate hands.

The unsuccessful attempt at scuttling enabled the Confederate Navy to raise and rebuild her as *Virginia*. *Virginia* was later sunk by the Confederates after her famous battle with USS Monitor to keep her from being taken back.

### SS *Kaiser Wilhelm der Grosse* (1914)

In August 1914, SS *Kaiser Wilhelm der Grosse* was requisitioned by the Kaiserliche Marine and converted into an auxiliary cruiser, assigned to commerce raiding in the Atlantic.

She was fitted with six 10.5 cm (4 inch) guns and two 37 mm guns. After sparing two passenger ships because they were carrying many women and children, she sank two freighters before she herself was sunk on 26 August 1914. She was caught refuelling off the shore of the then Spanish colony of Rio de Oro in western Africa by the old British 6-inch gunned cruiser HMS *Highflyer*.

Badly outgunned, the ship eventually ran out of ammunition. The crew abandoned and scuttled her. British sources at the time insisted that *Kaiser Wilhelm der Grosse* sank because of the damage inflicted by *Highflyer*.

### SMS *Dresden* (1915)

In December 1914, the *SMS Dresden* was the only German warship to escape destruction in the Battle of the Falkland Islands. She eluded her British pursuers for several more months, until she put into Más a Tierra in March 1915. Her engines were worn out and she had almost no coal left for her boilers. There, she was trapped by British cruisers, who violated Chilean neutrality and opened fire on the ship. The Dresden's Executive Officer - the future Admiral Wilhelm Canaris - negotiated with the British and bought time for his shipmates to scuttle the *Dresden*.

### Zeebrugge Raid (1918)

The Zeebrugge Raid involved three outdated British cruisers chosen to serve as blockships in the German-held Belgian port of Bruges-Zeebrugge from which German U-boat operations threatened British shipping. *Thetis, Intrepid* and *Iphigenia* were filled with concrete then sent to block a critical canal. Heavy defensive fire caused the *Thetis* to scuttle prematurely; the other two cruisers sank themselves successfully in the narrowest part of the canal.

Within three days, however, the Germans had broken through the western bank of the canal to create a shallow detour for their submarines to move past the blockships at high tide.

**German fleet at Scapa Flow (1919)**

In 1919, over 50 warships of the German High Seas Fleet were scuttled by their crews at Scapa Flow following the deliverance of the fleet as part of the terms of the German surrender. Rear Admiral Ludwig von Reuter ordered the sinkings, denying the majority of the ships to the British.

Von Reuter was made a prisoner-of-war in Britain but his defiant final act of war was celebrated in Germany. Though most of the fleet was subsequently salvaged by engineer Ernest Cox, a number of warships (including three battleships) remain, making the area very popular amongst undersea diving enthusiasts.

***Admiral Graf Spee* (1939)**

Following the Battle of the River Plate the damaged German pocket battleship *Admiral Graf Spee* sought refuge in the port of Montevideo. On 17 December 1939, with the British and Commonwealth cruisers HMS *Ajax*, HMS *Cumberland*, and HMNZS *Achilles* waiting in international waters outside the mouth of the Río de la Plata, the *Graf Spee* sailed just outside the harbour and was scuttled by Captain Hans Langsdorff to avoid risking the lives of his crew in what he expected to be a losing battle. Langsdorff shot himself three days later.

**Blockade of Massawa (1941)**

As the British advanced toward Eritrea during their East African Campaign in World War II, Mario Bonetti, the Italian commander of the Red Sea Flotilla based at Massawa, realized that his harbor was going to be overrun by the enemy.

In the first week of April, 1941, he began to destroy the harbor's facilities and ruin its usefulness to the British. Bonetti ordered the sinking of two large floating dry docks and supervised the calculated scuttling of eighteen large commercial ships in the mouths of the north Naval Harbor, the central Commercial Harbor and the main South Harbor, blocking navigation in and out. Scuttled, too, was a large floating crane.

The harbor was rendered useless by 8 April 1941 when it was surrendered to the British. Scuttled ships included the German steamers *Liebenfels, Frauenfels, Lichtenfels, Crefeld, Gera* and *Oliva*. Also scuttled were the Italian steamers *Adua, Brenta, Arabia, Romolo Gessi, Vesuvio, XXIII Marzo, Antonia C., Riva Ligure, Clelia Campenella, Prometeo* and the Italian tanker *Giove*.

The largest scuttled vessel was the 11,760 ton *Colombo*, an Italian steamer. The most dangerous problems for salvage were the *Brenta* which contained a booby trap sunk in one hold made of an armed naval mine sitting on three torpedo warheads and Regia Marina minelayer *Ostia* which had been sunk by the RAF with several of its mines still racked.

Thirteen additional coastal steamers and small naval vessels were scuttled as well. Though a civilian contractor was retained to clear a navigable passage through the wrecks, it wasn't until a year later that headway was made in the effort to return Massawa to military duties. U.S. Navy Commander Edward Ellsberg arrived in April, 1942 with a salvage crew and a small collection of specialized tools and began methodically correcting the damage. His salvage efforts were to yield significant results in just 5½ weeks.

On 8 May 1942, the SS *Koritza*, an armed Greek steamer, was drydocked for cleaning and minor hull repairs. Massawa's first major surface fleet 'customer' was HMS *Dido* which needed repairs to a heavily damaged stern in mid-August, 1942. Many of the harbor's sunken ships were patched by divers, refloated, repaired and taken into service. The *Ostia* and *Brenta* were successfully salvaged despite their armed mines.

### *Bismarck* (1941)

In 1941, the battleship *Bismarck*, heavily damaged by the Royal Navy, leaking fuel, listing, rendered mostly unmanoeuvreable and with no effective weapons but still afloat, was reported to have been scuttled by her crew to avoid capture.

This was supported by survivors' reports in *Pursuit: the Sinking of the Bismarck*, by Ludovic Kennedy, 1974 and by a later examination of the wreck itself by Dr. Robert Ballard in 1989. A later, more advanced examination found torpedoes had penetrated the second deck, normally always above water and only possible on an already sinking ship, thus further supporting that the scuttling had made the final torpedoing redundant.

### Coral Sea and Midway (1942)

After the Battles of the Coral Sea and Midway, the heavily damaged American aircraft carrier *Lexington* and the Japanese ones *Hiryû*, *Sôryû*, *Akagi*, and *Kaga* were all scuttled to prevent their preservation and use by their respective enemies.

### The French fleet in Toulon (1942)

In November 1942, in an operation codenamed *Case Anton*, Nazi German forces occupied the so-called "Free Zone" in response to the Allied landing in North Africa. On 27 November they reached Toulon, where the majority of the French Navy was anchored. To avoid capture by the Nazis (Operation *Lila*), the French admirals-in-command (Laborde and Marquis) decided to scuttle the 230,000 tonne fleet, most notably, the battleships *Dunkerque* and *Strasbourg*. 80% of the fleet was utterly destroyed, all of the capital ships proving impossible to repair. Legally, the scuttling of the fleet was allowed under the terms of the 1940 Armistice with Germany.

### The Danish fleet (1943)

*"The Danish fleet has been sunk with honour. Long live the Danish Fleet."* (speech by Commander Ipsen, translated)

Anticipating a German seizure of all units of the Danish Navy, mostly in Copenhagen but also at other harbours and at sea in Danish waters, the Danish Admiralty had instructed its captains to resist, short of outright fighting, any German attempts to assume control over their vessels, by scuttling if escape to Sweden was not possible. Preparations were made under the operational codeword "Safari".

Of the fifty-two vessels in the Danish Navy on 29 August, two were in Greenland, thirty-two were scuttled, four reached Sweden and fourteen were taken undamaged by the Germans. Nine Danish sailors lost their lives and ten were wounded.

Subsequently major parts of the Naval personnel were interned for a period. Just after, on the evening of 29 August, Admiral Wurmbach, Supreme Commander of the German Kriegsmarine in Denmark addressed Vice Admiral A H Vedel, the Commander in Chief of the Royal Danish Navy, as follows "Wir haben beide unsere Pflicht getan" (We have both done our duty).

### Allied landing in Normandy (1944)

Old ships code-named "Corn cobs" were sunk to form a protective reef for the Mulberry harbours at Arromanches and Omaha Beach for the Normandy landings. The sheltered waters created by these scuttled ships were called "Gooseberries" and protected the harbours so transport ships could unload without being hampered by waves.

## CONTEMPORARY ERA

Today, ships (and other objects of similar size) are sometimes sunk in order to help the formation of artificial reefs, as was done with the former USS *Oriskany* in 2006.

It is also common for military organizations to use old ships as targets, in war games, or for various other experiments. As an example, the decommissioned aircraft carrier USS *America* was subjected to surface and underwater explosions in 2005 as part of classified research to help design the next generation of carriers (the CVN-21 class), before being sunk with demolition charges.

Ships are increasingly being scuttled as a method of disposal. The economic benefit of scuttling a ship includes removal of ongoing operational expense to keep the vessel seaworthy. Controversy surrounds the practice. Notable actions against the practice include the USS *Oriskany*, which was scuttled with 700 pounds of PCBs remaining on board as a component in cable insulation, contravening the Stockholm Convention on safe disposal of persistent organic pollutants which has a zero tolerance for PCB dumping in marine environments.

The planned scuttling of the Australian frigate HMAS *Adelaide* at Avoca Beach, New South Wales in March 2010 was placed on hold after resident action groups aired concerns about possible impact on the area's tides and that the removal of dangerous substances from the ship was not thorough enough. Further cleanup work on the hulk was ordered, and despite further attempts to delay, *Adelaide* was scuttled on 13 April 2011.

Scuttled ships have been used as conveyance for dangerous materials. In the late 1960s, the United States Army scuttled the SS *Corporal Eric G. Gibson* and SS *Mormactern* with VX nerve gas rockets aboard as part of Operation CHASE – "CHASE" being Pentagon shorthand for "Cut Holes and Sink 'Em." Other ships have been "chased" containing mustard agents, bombs, land mines, and radioactive waste.

During the Falklands War, the Argentine submarine ARA *Santa Fe* was attacked and damaged by British helicopters on 23 April 1982. The crew abandoned the submarine on South Georgia and were captured by British forces. The submarine was later scuttled by the British.

The Iranian converted minelayer *Iran Ajr* was caught laying mines in the Persian Gulf during the Iran–Iraq War on 21 September 1987, in an effort to sink or damage U.S.-flagged Kuwaiti oil tankers in international waters. The ship was attacked by U.S. Army helicopter gunships flying from a U.S. Navy frigate, then boarded by SEALs, who captured the surviving crew, confirmed the presence of mines, and then scuttled the ship five days later.

In Somalian waters, pirate ships captured are scuttled. Most nations have little interest in prosecuting the pirates, thus this is usually the only repercussion.

**Narco submarines**

Narco submarines are known to be used by Colombian drug cartel members to import cocaine from Colombia to the United States. Due to their fiberglass construction, they are nearly invisible to radar, sonar, and infrared. After unloading or when intercepted, the submersible is scuttled; with no evidence of trafficking against them, the crewmen go from suspected traffickers to castaways who, in accordance with maritime law, must be rescued and cannot be charged.

However, laws were recently changed to address what was seen as an exploitation of legal loopholes. The United States now considers the operation of an unflagged (unregistered) vessel designed solely for the clandestine transport of contraband a crime in and of itself that carries severe legal penalties.

# RIGS-TO-REEFS

Rigs-to-Reefs (RTR) is a nationwide program in the United States developed by the former Minerals Management Service (MMS), now Bureau of Safety and Environmental Enforcement (BSEE), of the U.S. Department of the Interior to turn decommissioned offshore oil and petroleum rigs into artificial reefs.

The United States began extracting oil offshore in the early 20th Century; yet the concern over abandoned oil rigs surfaced only in the 1980s. "Today over 4,500 offshore oil and gas platforms have been installed supplying 25% of the United States' production of natural gas and 10% of its oil." These platforms continue to function as long as the reservoirs underneath them provide oil at a profitable rate. At the end of their productive lives they must be decommissioned and removed within one year.

An alternative to removal is to turn the rig into a reef through the Rigs-to-Reef (RTR) program.

All coastal states have such artificial reef programs in the interest of increasing ocean fisheries but not all participate in RTR. The rig's steel structures are stable and durable. They create shelter for marine life in open waters where there was none.

## PROCESS

Once a rig stops producing economic quantities of oil, the site is usually abandoned. The Minerals Management Service (MMS) requires the operator to remove the rig within a year of abandonment (stopped production) and lease end. MMS supports and encourages RTR as an alternative to total removal.

RTR recognizes that during a rig's productive years, significant marine life comes to live on and around its structure. RTR preserves much of that marine life and encourages further growth. The operator benefits by avoiding the substantial cost of removal. Cumulative costs of removal had reached an estimated $1 billion by the year 2000. The shape and complexity of the structure may lead to significant species diversity.

### Decommissioning a platform

Officially, decommissioning an oil rig is the act of removal according to regulatory requirements and includes flushing, plugging and cementing wells to make them safe. Decommissioning is complicated by factors such as cost, safety, operational duration, environmental issues, risk, experience, and historical relationship between operator and state. As part of decommissioning, the operator must deal with the shell mound that collects on the bottom surrounding the rig. The mound consists of shells that have fallen from the rig, material that has fallen and/or leaked from the platform, mixed with well seepage.

Mounds can contain significant levels of toxic metals including, arsenic, cadmium, chromium, copper, nickel, PCBs, lead, zinc, and poly-nuclear hydrocarbons. Removing the rig structure does not eliminate the need to address the mound.

The method of decommissioning depends on water depth and structure type and is a three step process that includes planning, permitting, and implementation. A party other than the operator usually administers the process.

In Louisiana, costs as well as the risk involved are the primary factors in determining how to decommission rigs. If the savings are large enough, the operator typically chooses reefing and donates 1/2 the savings to maintain the reef. Decommissioning a shallow water rig typically costs $10–15 million so the amounts can be substantial.

The Louisiana Artificial Reef program from its inception through 1998 received roughly $9.7 million in donations and has not taken taxpayer money.

**Methods of reefing**

Severing the rig from the bottom using explosives is the easiest approach, but has the potential to harm marine life. This potential is greatly reduced if the explosives are all placed deep below the seafloor. Current requirements place the explosives a minimum of 5 metres (16 ft) below the seafloor which eliminates the threat to all but the closest sea turtles.

National Oceanographic and Atmospheric Administration (NOAA) National Marine Fisheries Service (NMFS) marine observers and helicopter surveys hours preceding the event keep most sea turtles away from the area.

Alternatively, commercial divers can use mechanical and abrasive cutters, which preserves marine life, but places the divers at considerable risk.

Reefing involves one of three methods.:

- Tow and Place: Sever the structure from the sea floor and tow it to a state-approved location.
- Partial Removal: Remove the top portion of the submerged platform and either place it on the sea floor or tow it to another site. Partial removal can result in a loss of the shell mound community and fish that live in the top section but the rest (the majority) of the habitat remains intact.
- Toppling: Toppling involves uses explosives to sever the base of the structure below the mud line. Then it simply falls over. Toppling eliminates shallow and mid-ocean habitats. However, these portions of the rig are quickly occupied by other creatures.

## HISTORY

In the early 1900s coastal oil exploration and drilling began around the

United States. Initially, no laws governed this activity. In 1979, Exxon relocated their experimental subsea production system from offshore Louisiana to a permitted artificial reef site off Apalachicola, Florida. The first platform jacket was donated by Tenneco and towed from Louisiana to Pensacola, Florida.

By 2000, companies had donated 141 platforms to the effort.

## LEGISLATION

Congress passed the Submerged Lands Act in 1953 to clarify responsibilities between state and federal government. This was immediately followed by the Outer Continental Shelf Lands Act (OCSLA) used to control leasing of exploration rights in the Outer Continental Shelf (OCS).

The OCSLA did not contain any real environmental provisions associated with drilling and in 1969 an explosion at a drilling location off of California's coast triggered the National Environmental Policy Act (NEPA).

NEPA concluded that every major federal action (i.e.: oil exploration on the OCS) required an Environmental Impact Statement (EIS). In 1982, The U.S. Department of the Interior created the Mineral Management Service (MMS) to monitor development on the Outer Continental Shelf. The MMS leases submerged federal lands and assesses the environmental effects of exploration and drilling (by issuing an EIS).

In 1984 Congress passed the National Fishing Enhancement Act (NFEA) which provided the basis for artificial reef programs. The NFEA spawned the National Artificial Reef Plan of 1985. This plan cleared the way for government-endorsed artificial reef projects and subsequently the Minerals Management Services' Rigs-to-Reef program.

In California, a bill granting oil companies the right to conditional partial removal was approved and signed by Governor Arnold Schwarzenegger in 2010. A.B. 2503 "allows a platform owner or operator to design a 'partial removal' plan for a platform and to apply for permission to implement it". The plan requires the approval of three agencies: The Department of Fish and Game (DFG), The Department of Ocean Protection Council (OPC), and the California State Lands Commission.

## TRENDS

Rigs-to-Reefs was first explored in 1979 when the first oil rig was transported from Louisiana to a Florida site. This rig was the first of 5 Rigs-to-Reefs towed to Florida's coast. Louisiana was the first state to develop a program that allowed transfer of liability and ownership from the operator to the state.

Texas later followed this example. Rigs-to-Reef is now the core of both Louisiana and Texas' artificial reef programs. From 1987-1995, of the over

941 platforms removed from Louisiana and Texas waters, 90 became artificial reefs.

By November 2000, 151 platforms had been recycled as artificial reefs in the Gulf of Mexico from states such as Texas, Louisiana, Mississippi, Alabama, and Florida. Federal law now requires rigs to be demolished for safety reasons, although discussion continues about their potential.

## DEBATES

Opposition, and also support, for Rigs-to-Reefs came from environmentalists, fisherman, oil companies, and others. California and the North Sea are each debating RTR.

In California, legislation was proposed during the 2010 session to clear legal hurdles for RTR; the Coastal Commission held hearings, but then the legislation was tabled. Even with the RTR successes in the Gulf of Mexico and Philippines, differences in terrain, government entities, and concerned citizens generated conflict.

### Fisheries

A 2000 MMS report lists research that shows fish densities 20 to 50 times higher around oil and gas platforms than in nearby open water. Divers assess fish populations surrounding platforms.

The report encourages recreational fisherman, divers and others who benefit from the increased density. Opponents claim that the greater density comes from an influx of nearby fish rather than increased total population. Research observing rockfish populations on oil rigs supports both theories.

### Liability

Commercial fisherman who trawl sometimes oppose RTR because their nets may snag a rig, creating a hazardous situation. Several fisherman have reported tangling their nets on submerged rigs. Navigational mishaps and diving accidents may also occur around an artificial reef. Gulf of Mexico Rigs-to-Reefs participants have not yet reported any liability problems.

The MMS does not release an operator from liability unless another entity accepts ongoing liability for the rig. If the reef is in state waters, the state typically accepts liability. In federal waters, liability typically goes to a private entity or to another MMS-approved agency. Critics claim that the primary reason that operators support RTR is their desire to offload decommissioning costs and liability.

Liability is divisible, however. In 2001, the California legislature passed, although the governor then vetoed, a bill that would allow operators to transfer liability to another entity, while retaining liability for any pollution from the underlying well.

**Ethics**

As with cap-and-trade and ecotourism, RTR attempts to enlist the private sector in helping the environment. In the minds of many environmentalists, such efforts are fundamentally suspect. Environmental groups have long opposed oil companies and frame their critique around distrusting these parties.

"No other industry is allowed to leave a toxic mess for the state to manage and maintain at taxpayer expense" said Linda Krop, Chief council for the Santa Barbara based Environmental Defense Center.

## OPERATION REGENBOGEN (U-BOAT)

Regenbogen ("Rainbow") was the code name for the planned mass scuttling of the German U-boat fleet, to avoid surrender, at the end of World War II.

### BACKGROUND

At the beginning of May 1945 Nazi Germany was collapsing under the Allied onslaught.

The Soviets had captured Berlin, and on 30 April Hitler had committed suicide. He had appointed Grand Admiral Karl Dönitz as Head of State and Supreme Commander of the Armed Forces. US Forces from the west and Soviet forces from the east had already met at Torgau, cutting the Reich in two, while in the north 21st Army Group was poised to capture Hamburg and the other German ports.

The state of the German navy, the *Kriegsmarine,* was no better. Of its capital ships only *Prinz Eugen* survived, sheltering at Copenhagen; only the U-Boat Arm was capable of continuing the fight. The *Kriegsmarine* had approximately 470 U-boats remaining.

Some 170 of these were operational U-boats (*Front* boats), based mainly in occupied Norway, and another 200 home-based boats in various stages of building, commissioning and working up; these were mainly in the north German ports and on the Baltic.

As head of the *Kriegsmarine,* and as commander of the U-boat arm, Dönitz was keen that his U-boat force should not be surrendered. However as the new German leader, he was keen to extricate Germany from the war and, if possible, avoid Allied, particularly Soviet, retribution. To that end he had opened negotiations with the western Allies, through Field Marshal Montgomery commander of Allied 21 Army Group, in North Germany.

As the Allies closed in on the North German ports the *Kriegsmarine* started to destroy what was left to prevent its capture, while all servicable boats were ordered to bases in Norway. During May a final massacre of U-boats fleeing to Norway took place; 23 U-boats were destroyed or damaged beyond repair in transit in the first week of May.

Against this backdrop, Dönitz and the U-boat arm made plans for a mass scuttle of his U-boats, to be carried out on receiving the code-word "Regenbogen".

## THE OPERATION

During May the *Kriegsmarine* started to scuttle its U-boats ahead of the advancing Allied armies.

On 1 May 1945, 3 U-boats were wrecked at Warnemünde, outside Rostock on the Baltic coast, the first of a wave of scuttling boats and destroying facilities.

On 2 May, a further 32 U-boats were scuttled at Travemünde, near Lübeck. On 3 May, Dönitz sent his chief aide von Friedeburg to Montgomery at Lüneburg to open negotiations for an armistice with the western allies. This was refused, as von Friedeburg was not empowered to agree to an unconditional surrender, upon which the Allies insisted. Also on 3 May another 39 boats were wrecked, 32 at Kiel and another 7 at Hamburg, on the North Sea coast.

On 4 May the Supreme Allied Commander, Eisenhower allowed the German forces in Northwest Europe, including naval forces, to surrender to Montgomery and 21 Group; for this, Montgomery insisted that German naval forces, including the U-Boat arm, be surrendered intact. This would come into effect at 8 am on 5 May. Meanwhile on 4 May, four more boats, two in the Kiel Canal and two at Flensburg, were scuttled.

In the early hours of 5 May, the *Regenbogen* order was given, only to be countermanded 8 minutes later, to avoid jeopardizing the surrender negotiations, and later that day all operational U-boats were ordered to cease hostilities.

Despite this a further 87 boats were destroyed on 5 May; 64 on the Baltic (41 at Gelting Bay, 13 at Flensburg and 10 at various other points), while on the North Sea coast 23 boats were disposed of, 13 at Wilhelmshaven and 10 in the Weser estuary.

On 6 May, there were no further sinkings, but on 7 May the two Walther boats were wrecked at Cuxhaven. Over the last week at least 195 U-boats had been scuttled.

On 8 May, Germany surrendered unconditionally; the remaining naval units, including the surviving U-boats, surrendered to Allied forces. At least 150 U-boats were surrendered to the Allied navies, either at sea or at their operational bases. 52 boats were surrendered at sea, either on patrol or in transit, and 98 in port, mostly in Norway and at bases in Germany, Denmark and France.

Four U-boats and their crews fled to neutral ports rather than do this; two (*U-1277* and *U-963*) to Portugal and two (*U-530* and *U-977*) to Argentina.

## DECOMMISSIONING OF RUSSIAN NUCLEAR-POWERED VESSELS

The decommissioning of Russian nuclear-powered vessels is an issue of major concern to the United States and to the Scandinavian countries near Russia.

From 1950 to 2003, the Soviet Union and its major successor state, Russia, constructed the largest nuclear-powered navy in the world, more ships than all other navies combined: 248 submarines (91 attack submarines, 62 cruise missile submarines, 91 ballistic missile submarines and four research submarines), four Kirov class battlecruisers, and a missile test ship, as well as nine icebreakers. Many were or are powered by two reactors each, bringing the total to 468 reactors.

With the end of the Cold War and with its navy chronically underfunded, Russia has decommissioned many of these vessels, and according to one report dated November 2008, intends to scrap all decommissioned submarines (more than 200) by 2012. However, the safety records of the Soviet and Russian navies and the budgetary constraints on the Russian government are matters of great concern.

Ships awaiting decommissioning receive little maintenance, and there are insufficient waste storage facilities, raising worries about possible ecological damage from accidents or improper storage.

In the midst of the Cold War, the Soviet Union was more concerned about building up its fleet, rather than making provisions for the disposal of aging nuclear-powered vessels. It was not until 1986 that "the Central Committee of the Communist party and the Supreme Soviet ratified Decree No. 095-296 which laid down formal procedures for decommissioning and dismantling inactive nuclear submarines."

By this time, the first generation November class submarines had been in service for over 20 years (by comparison, the service life expectancy of the more modern Delta IVs is estimated to be between 20 and 30 years with regular overhauls, or 10 to 15 without).

Progress was hindered by the large number of government bodies involved, resulting in much confusion and competition. The breakup of the Soviet Union in 1991 further complicated matters. Russia assumed responsibility for all of the Soviet Navy's nuclear-powered ships, but agreements between agencies responsible for decommissioning had to be renegotiated.

The main problem is a lack of money. In 1995, a Northern Fleet submarine based near Murmansk nearly suffered a nuclear meltdown when power was cut off due to unpaid electricity bills. Decommissioned vessels are often left in floating storage until funds can be allocated for their dismantling. As of November 2001, "up to 40% of the decommissioned submarines have been in floating storage without much maintenance for more than 10 years".

The situation has caused so much concern to other countries, in particular the United States, Japan and the Scandinavian countries, that they have contributed funds and other assistance.

For example, the Arctic Military Environmental Cooperation (AMEC) was a joint Norwegian, Russian, and American government consortium (later joined by the British) set up to deal with military environmental issues, mainly the dismantling of Russia's nuclear submarine fleet in Europe. After the "somewhat acrimonious dissolution" of AMEC, the Norwegian and British governments shared the £3.9 million cost of dismantling a November class submarine.

Under AMEC's successor, Cooperative Threat Reduction, the British government financed the dismantling of two Oscar I submarines. The Nunn–Lugar Cooperative Threat Reduction Program has been responsible for the deactivation and destruction of many weapons, including 33 nuclear submarines. With the "Star of Hope" program, Japan has funded the dismantling of five Victor III and one Charlie I submarines in the Far East.

Another problem is security. Russian sailors have been convicted and jailed for two 1993 thefts of highly enriched uranium from fuel rods. In 1994, Russian officials caught two North Korean agents trying to buy submarine dismantlement schedules.

## REACTOR DISPOSAL

Removing the fuel from a nuclear reactor requires a specially trained team. The coolant is drained first. A reactor must be cooled down for at least three years after its final shutdown before this can be done. The hull above the reactor is then removed, followed by the top shield. The fuel elements are extracted and transported by ship and then rail to a storage facility.

The still heavily radioactive reactor compartment can then be cut away. (Most Russian submarines have two reactors, "in separate rooms, but in the same compartment.") Due to a lack of land storage facilities, two adjoining emptied compartments, one forward and one aft, are usually cut off as well to provide buoyancy for storage on water.

In some cases, however, only the reactor compartment is removed, and pontoons attached to keep it afloat. A third method involves filling the reactor compartment with polisterol for buoyancy. "Reactor compartments from Polyarny and other shipyards at the Kola Peninsula and in Severodvinsk, Arkhangelsk county, are towed to Sayda Bay".

On 10 August 1985, control rods were incorrectly removed from a Victor class submarine during defueling at Chazma Bay naval yard outside Vladivostok, resulting in an explosion, the "release of large amounts of radioactivity", and ten deaths. In addition to this Victor, five other submarines have damaged cores preventing their defueling by normal methods.

A Russian government report acknowledged in March 1993, that "during the period of 1965 to 1988 the Northern Fleet had dumped four reactor compartments with eight reactors (three containing damaged fuel) in the Abrosimov Gulf in 20 to 40 meters of water." Six other compartments, containing nine reactors in all, had also been dumped into the water in the 1960s and 1970s.

## SUBMARINES

### November class

The surviving November class submarines were decommissioned between 1986 and 1990. Several of them have been scrapped already. All of the survivors remain laid-up hulks in Russian naval bases (*K-14*, *K-42*, *K-115* and *K-133* of the Pacific Fleet; *K-11* and *K-21* of the Northern Fleet). There are plans to convert the first submarine of the class (*K-3*) into a museum ship in St. Petersburg, but the hulk remains in Polyarny, Murmansk Oblast, due to economic reasons and the "radiophobia" of some ecological organizations.

*K-5* was defueled at the naval yard at Polyarny in November 1966. According to one source, the shipyard "likely dismantled" the submarine the same year.

*K-27* was an experimental attack submarine of the November class that went into service in October 1963. A reactor malfunction on 24 May 1968 resulted in the release of radioactive gas into the engine room and fatal exposures to nine crewmen.

Replacing the reactor was considered and discarded. *K-27* was officially decommissioned on 1 February 1979. "The empty spaces of the reactor and equipment associated with the reactor... were filled with a solidifying radiation-resistant preservative" and she was scuttled off the northeastern coast of Novaya Zemlya on 6 September 1982.

*K-159* sank in the Barents Sea on 30 August 2003, while being towed to be scrapped, killing nine crewmen and depositing 800 kilograms (1,800 lb) of reactor fuel on the seabed.

### Hotel class

All eight Hotel class submarines were decommissioned for scrapping between 1987 and 1991. One undated article placed one at Sevmorput Naval Shipyard No. 35 in Murmansk. As of February 2003, two were reported to be at Russian Shipyard Number 10 at Polyarny.

### Echo class

The 34 Echo class submarines were decommissioned for scrapping between 1985 and 1995. According to the Federation of American Scientists, one or possibly two Echo Is were decommissioned in the mid-1980s and the remaining three or four in 1990. The 29 Echo IIs were all decommissioned by the mid-1990s.

On 23 October 2002, a nuclear submarine caught fire during dismantlement at Sevmorput; according to the Bellona Foundation, this was probably the Echo II *K-22*. Fortunately, the reactor compartment had already been removed, and there was no danger of radioactive contamination.

**Papa class**

On 4 June 2010, Voice of Russia reported that *K-222*, the sole member of the Papa class, had been scrapped at Severodvinsk.

**Yankee class**

As a result of the SALT I and START I/II treaties, 33 of the 34 boats of the Yankee class have been decommissioned. *K-219* was lost on 6 October 1986 after an explosion and fire near Bermuda. The Bellona Foundation reported on 7 February 2003 that:

- *K-214* had been dismantled at Sevmash shipyard "as an experiment to determine whether the yard can handle such operations."
- ten were dismantled at Zvezdochka shipyard: *K-216, K-415, K-137, K-140, K-210, K-228, K-444, K-221, K-418*, and *K-32*

**Delta class**

All 18 Delta I boats were taken out of service by 1998 and scrapped by 2005, while the four Delta IIs were taken out of service in 1996. Some of the Delta IIIs have been decommissioned and others put in reserve. Five remain in active service, along with all seven Delta IVs. Delta Is *K-279, K-385, K-472* and *K-475*, Delta II *K-193*, and Delta IIIs *K-441* and *K-424* were reported dismantled at Zvezdochka shipyard.

**Charlie class**

The 11 Charlie Is and seven Charlie IIs were decommissioned for scrapping between 1990 and 1994. On 4 June 1997, a Charlie "sank in a harbor on the Kamchatka Peninsula".

However, a local official reported that the reactor core had already been removed. On 21 November 2005, Russian President Vladimir Putin and Japanese Prime Minister Junichiro Koizumi reached an agreement to dismantle five decommissioned nuclear submarines, among them a Charlie I, the latter "the first project in Kamchatka under the Japan-Russia cooperation framework."

**Victor class**

According to some sources, all 16 Victor Is and seven Victor IIs were decommissioned by 1996. One cited "various sources" in estimating that "somewhere between 8 and 15 [Victor IIIs] had been decommissioned due to lack of funds in the 1999-2000 timeframe" while another placed the number at "about a dozen" (of the 26 built) by 1996.

In February 2003, under the Russian-Japanese "Star of Hope" program, Japan spent $6 million to dismantle a Victor III of Russia's Pacific Fleet; the job was completed in December 2004. Another 20 billion yen ($171 million) was allocated to dismantle four Victor IIIs at the Zvezda shipyard in the town of Bolshoi Kamen and a Charlie I at Vilyuchinsk, Kamchatka (as previously mentioned).

**Alfa class**

*K-64,* the first of seven Alfa class submarines, joined the Soviet Navy in December 1971. The following year, however, "the *K-64* suffered a major reactor problem when the liquid metal in the primary coolant hardened". Her hull was cut in half in 1973-1974 at Sverodvinsk, the forward portion being sent to Leningrad to be used for training, the reactor compartment being kept at Zvezdochka. The remaining boats were decommissioned for scrapping, five in 1990 and one in 1996.

As of February 2003, *K-463, K-316, K-432* and *K-493* had been dismantled at Sevmash, with the remaining two at Bolshaya Lopatka, Zapadnaya Litsa, awaiting the same fate.

The Alfa reactors use a liquid metal coolant that must be kept at a temperature over 123°C to avoid solidifying; a 2002 paper reports that the coolant in both *K-123* and *K-373* has solidified. An Alfa reactor compartment was reported in November 1997 as being at "an open site on Yagry island in Severodvinsk".

**Oscar class**

The two Oscar Is were decommissioned in 1996. By 2006, it was reported that their dismantling, funded by the United Kingdom under the Cooperative Threat Reduction program, had been completed at Sevmash shipyard. Two or possibly three Oscar IIs "were inactivated in the late 1990s, and as of mid-2000 were laid up awaiting disposal."

**Typhoon class**

Three of the six Typhoons have been scrapped. *TK-202* was deactivated in 1997, and was laid up from July 1999 at the Sevmash shipyard in Severodvinsk.

With funding from the Nunn–Lugar Cooperative Threat Reduction program, the defueling of her reactors started in June 2002 at the Zvezdochka shipyard. Scrapping was completed in 2003-2005. *TK-12* was withdrawn from active service in 1996, and scrapped in 2006-2008. *TK-13* was withdrawn from active service in 1997, and scrapped in 2007-2009, "funded by Russia, USA and Canada as part of the program 'Global partnership'". In September 2011, the Russian defense ministry announced it will decommission and dismantle the three remaining boats to comply with the START III treaty and because the newer Borei class submarines require smaller crews and are less costly to maintain.

**Akula class**

*K-284 Akula,* the lead ship of the class, "was decommissioned in 1995 to avoid the expense of a reactor refueling".

## SURFACE SHIPS

NS *Lenin,* the world's first nuclear-powered surface and civilian ship, was decommissioned in 1989. It was subsequently converted to a museum ship.

Of the six Arktika class icebreakers, two are no longer in service. NS *Arktika* was taken out of service and had its reactor removed in 2008. NS *Sibir* was taken out of service in 1992 due to a problem in the vessel's steam generation system.

The *Barents Observer* reported on 3 November 2008 that the Russian authorities were preparing to decommission and scrap *Sibir*. In 2010, the deputy chief engineer of Atomflot, the Russian nuclear fleet operator, reported that "all radioactive [materials] have been unloaded" from *Siber*, but the "decommissioning decision has not been made yet, however."

# ARTIFICIAL REEF

An artificial reef is a human-made underwater structure, typically built to promote marine life in areas with a generally featureless bottom, control erosion, block ship passage, or improve surfing.

Many reefs are built using objects that were built for other purposes, for example by sinking oil rigs (through the Rigs-to-Reefs program), scuttling ships, or by deploying rubble or construction debris. Other artificial reefs are purpose built (e.g. the reef balls) from PVC or concrete. Shipwrecks may become artificial reefs when preserved on the sea floor. Regardless of construction method, artificial reefs generally provide hard surfaces where algae and invertebrates such as barnacles, corals, and oysters attach; the accumulation of attached marine life in turn provides intricate structure and food for assemblages of fish.

## HISTORY

The construction of artificial reefs is thousands of years old. Ancient Persians blocked the mouth of the Tigris River to thwart Indian pirates by building an artificial reef, and during the First Punic War the Romans built a reef across the mouth of the Carthaginian harbor in Sicily to trap the enemy ships within and assist in driving the Carthaginians from the island. Artificial reefs to increase fish yields or for algaculture have been used at least since 17th century Japan, when rubble and rocks were used to grow kelp, while the earliest recorded construction of artificial reef in the United States is from the 1830s when logs from huts were used off the coast of South Carolina to improve fishing.

Since at least the 1830s, American fishermen used interlaced logs to build artificial reefs. More recently, castaway junk, such as old refrigerators, shopping carts, ditched cars, out-of-service vending machines replaced the logs in ad hoc reefs. Officially sanctioned projects have incorporated decommissioned subway cars, vintage battle tanks, armored personnel carriers and oil drilling rigs.

## DEVELOPMENT

Artificial reefs tend to develop in more or less predictable stages. First, where an ocean current encounters a vertical structure, it can create a plankton-rich upwelling that provides a reliable feeding spot for small fish such as sardines and minnows, which draw in pelagic predators like bluefin tuna and sharks. Next come creatures seeking protection from the ocean's lethal openness—hole and crevice dwellers such as grouper, snapper, squirrelfish, eels, and triggerfish. Opportunistic predators such as jack and barracuda also appear, waiting for their prey to venture out. Over months and years the reef structure becomes encrusted with algae, tunicates, hard and soft corals, and sponges.

## REEFS

### Florida

Florida is the site of many artificial reefs, many created from deliberately sunken ships, including Coast Guard cutters Duane and Bibb and the U.S. Navy landing ship Spiegel Grove.

### Osborne Reef

In the early 1970s, more than 2,000,000 used vehicle tires were dumped off the coast of Fort Lauderdale, Florida to form an artificial reef. Catastrophically, the tires were not properly secured to the reef structures. Ocean currents broke them loose, sending them crashing into the developing reef and its natural neighbors. As of 2009, less than 100,000 of the tires had been removed after more than 10 years of efforts.

### Neptune Reef

Neptune Reef was originally conceived as an art project that would gradually decay. Burial at sea became a way of financing the project. As of 2011, about 200 "placements" had taken place. Cremated remains are mixed with cement and either encased in columns or molded into sea-star, brain-coral, 15 foot-tall castings of lions or other shapes before entering the water.

**The ex-USS *Oriskany***

The world's largest artificial reef was created by the purposeful sinking of the aircraft carrier USS *Oriskany* off the coast of Pensacola, Florida, in 2006.

**The ex-USNS *Hoyt S. Vandenberg***

The second-largest artificial reef is the USNS *Hoyt S. Vandenberg*, a former World War II era troop transport that served as a spacecraft tracking ship after the war. The Vandenberg was scuttled seven miles off Key West on May 27, 2009, in 140 feet of clear water. Supporters expect the ship to draw recreational divers away from natural reefs, allowing those reefs to recover from damage from overuse.

**The ex-USS *Spiegel Grove***

The ex-Spiegel Grove is located on Dixie Shoal, 6 miles (10 km) off the Florida Keys in the Florida Keys National Marine Sanctuary. Her exact location is 25°04200.23N 80°18200.73W.

**Redbird Reef**

In late 2000, The MTA New York City Transit decided to phase out its outdated fleet of subway cars to make room for the R142 and R142A trains. These subway cars, (Redbirds), ran on the IRT lines in the New York City Subway system for 40 years. Each car was sold, stripped, decontaminated, loaded on a barge, and sunk in the Atlantic Ocean. Some had number plates removed because of rust, and auctioned off on eBay. 1,200 subway cars were sunk.

In September 2007 the MTA approved a contract worth $6 million, to send 1,600 of its retired subway cars to be used as artificial reefs. Most of these trains ran on The BMT/IND lines. The trains include the R32, R38, R40 and R42. These models are stainless steel. The MTA will replace them with the R160A and R160B trains. The plastic front ends were removed before sinking. The retired fleet included old work trains and cars damaged beyond repair.

**Cancun Underwater Museum**

Since November 2009, artist Jason Decaires Taylor has created more than 400 life size sculptures off the coast of Cancun, Mexico. The coral reefs in this region suffered heavy degradation due to repetitive hurricane abuse. This project funded by The National Marine Park and the Cancun Nautical. Association was designed to emmulate coral reefs using a neutral ph clay. Taylor has constructed unique settings depicting daily activities ranging from a man watching TV to a 1970's replica of a Volkswagen Beetle. This artificial reef has relieved pressure from the nearby Manchones Reef. The design and materials implemented in this project have proved the ecological viability of artificial reefs.

**Australia**

Since the late 1990s, the Australian government has been providing decommissioned warships for use as artificial reefs for recreational scuba diving. So far, the following six ships have been sunk:

- Ex-HMAS Swan at Dunsborough in Western Australia during December 1998.
- Ex-HMAS Perth at Albany in Western Australia during November 2001.
- Ex-HMAS Hobart in Yankalilla Bay in South Australia during November 2002.
- Ex-HMAS Brisbane off the Sunshine Coast in Queensland during July 2005.
- Ex-HMAS Canberra at a site west of the entrance to Port Phillip Bay in Victoria during October 2009.
- Ex-HMAS Adelaide off Terrigal on the New South Wales Central Coast during April 2011

**Gibraltar**

The Gibraltar Reef was first proposed by Eric Shaw in 1973. Initial experiments with tyres proved unsuccessful as the tyres were simply swept away by currents or buried underneath sand. In 1974, boats from local marinas and the Gibraltar Port Authority were donated.

The first two were barges that were sunk in Camp Bay. In 2006, a 65 ton wooden boat, "True Joy" (also referred to as "Noah's Ark") was sunk here as well, followed by the MV New Flame, a mid sized bulk carrier, in 2007.

In 2013, a dropping of more than 70 concrete blocks of one square meter with metal bars, took place. There was a discussion about, because some experts don't consider the blocks of concrete with metal bars a real reef but a system to bother Spanish fishing activities.

The blocks with metal devices proved unpopular with Spanish fishermen who have suffered important economic loses and demanded the process to be stopped because it would prevent them trawling the seabed for fish. The metal bars are designed to break the fishermen nets. The dropping lead to a diplomatic conflict between the Kingdom of Spain and United Kingdom.

## ARTIFICIAL SURFING REEFS

Artificial surfing reefs have been created for surfing, coastal protection, habitat enhancement and coastal research. The world's first attempt was made in El Segundo, near Los Angeles, in California. The next attempt was at Mosman Beach, Perth, Western Australia. This reef was constructed of large granite rocks placed in a pyramidal shape to form an appropriate breaking wave form that would suit surfers. An artificial reef constructed of over 400 massive, geotextile bags (each one larger than a bus) filled with sand was constructed in 2000 at Narrowneck on the Gold Coast of Queensland, Australia.

This artificial reef had two objectives: stabilizing beach nourishment and improving surfing conditions. In the United States, in particular, demanding coastal permit requirements and environmental opposition present major obstacles to building surfing reefs. As of February 2006, the only reef built in the U.S. for surfing is southern California's "Pratte's Reef", which failed to create waves. Reefs built to enhance marine habitat face less environmental opposition, in part because they are in deeper water and further offshore. A number of such man-made reefs exist near Florida and Hawaii. Artificial surfing reefs typically resemble a "submerged breakwater", and proponents suggest benefits beyond surfing conditions. Many coastlines are subject to powerful waves that crash directly onshore. An artificial reef 150-300 yards offshore might create surfing opportunities and, by dissipating wave energy, make swimming safer and reduce coastal erosion. The USS *Spiegel Grove* was sunk in 2002 to make an artificial reef. Europe's first artificial reef was approved in 2008.

Construction began August 30, 2008 in Boscombe, Bournemouth, UK, and opened in November 2009. The multi-purpose reef reef is expected to create waves up to 30% larger and double the number of surfing days annually. Construction on this reef began in June 2008, and was completed in August 2009. Boscombe Reef was built from large sand-filled geotextile containers, totaling 13,000 cubic meters.

## ELECTRO MINERAL ACCRETION (EMA)

Mineral accretion involves applying a low voltage current to a metallic structure to cause limestone to crystallize on the surface, to which coral planulae can attach and grow. The electric current also speeds post-attachment growth.

EMA works like charging a battery with a positive pole, the cathode, and a negative pole, the anode. Applying electric current attracts various dissolved minerals to either the cathode or the anode. Chemical reactions take place at both poles. On the anode, bubbles of oxygen and chlorine gas form. These bubbles float to the surface and dissolve into the air. On the cathode, bubbles of hydrogen gas and a limestone precipitate appear.

The voltage is low enough that it can be generated by floating solar panels or from wave motion. A coalition of scientists named the Global Coral Reef Alliance (GCRA) is developing a technique called the Biorock Process using mineral accretion for reef restoration, mariculture, and shoreline protection.

## ENVIRONMENTAL CONCERNS

According to The Ocean Conservancy, a Washington-based environmental group, the Osbourne reef may be an indication that the benefits of artificial reefs need to be re-examined. Jack Sobel, a senior scientist at the group, has said "There's little evidence that artificial reefs have a net benefit," citing concerns such as toxicity, damage to ecosystems and concentrating fish into one place (worsening overfishing).

# 6

# Ship Dismantling and Pre-cleaning of Ships

## INTRODUCTION

### SHIP RECYCLING AS A GLOBAL ACTIVITY

Dismantling of vessels was a common part of a port scene especially throughout southern Europe from the late 1940s to the 1960s. Lower labour costs and a domestic demand for steel meant that vessels were more likely to be scrapped in Japan, Korea and Taiwan during the late 1960s, 1970s and early 1980s, but for the last 25 years, the global centre for shipbreaking has been India, Bangladesh and Pakistan.

Here, the ships are scrapped directly on the beaches or what are in fact vast intertidal mudflats exposed daily by a 10 m tidal gauge. The beaching method of the Indian Subcontinent relies heavily on low labour cost since it involves very little mechanisation.

For more than 10 years shipbreaking has also been the issue of a strong public debate fuelled first by the images captured at Chittagong in Bangladesh by the Brazilian photographer Sebastiao Salgado and published in his book *"Workers"* from 1993.

Since then, the authors of a series of articles on the conditions in Alang, *"The Shipbreakers"*, were awarded the Pulitzer Prize in 1997, and not least the efforts of the NGOs, in particular the Greenpeace shipbreaking campaign, have regularly brought end-of-life ships, ship owners, shipbreaking yards and various authorities in the public spotlight and questioned the legality of the present practices.

The European Commission has addressed ship scrapping in several studies since 2000, and international bodies such as the International Labour Organisation (ILO), the Basel Convention Conference of the Parties and the International Maritime Organisation have increasingly addressed the issue of shipbreaking or ship recycling with the aim of bringing a global mandatory regime to the field.

Also, governments of EU member states and OECD countries have initiated actions to specifically regulate this issue. Lately, a few responsible ship owners have become reluctant to send their vessels to the scrapping beaches in Asia and are actively seeking alternatives.

In a much publicised event due to a decision of the Conseil d'État, the French government recalled the former French aircraft carrier 'Clemenceau' from its voyage to Alang, India, and it is now awaiting recycling in France under European standards of safety, health and environment. No active ship recycling facilities exist in the EU that can accept vessels the size of an aircraft carrier, and few can accommodate commercial vessels of quite common sizes such as Suezmax or Panamax.

The under-capacity and lack of special regulation and certification in 'green' recycling in Europe and elsewhere do not motivate today's ship owners to accept a loss on the recycling value on a vessel, when legal options exist for trading vessels near their end-oflife on a global market place.

In the present trading situation, commercial vessels are only reluctantly removed from trading since freight rates are high, and consequently, the scrapping volume in Asia in all vessel categories are also at a low point. Eventually, it must be expected that an accumulated volume of aging vessels will be obsolete and available for scrapping once the market cools off, in addition to the single-hulled oil tankers as prescribed in EU and IMO legislation.

After discussion in the context of the Basel Convention, ILO and IMO, and a Joint Working Group of the three organisations, IMO is working on a proposal for a new convention on ship recycling with the IMO as the lead organisation.

A first draft is available, but at present it may not adequately address several issues relevant to the Basel Convention and the EU Waste Shipment Regulation: Prior informed consent, the use of pre-cleaning, the requirements and certifications of the recycling facilities.

The ship owner organisations have, with some justification, claimed that the market for 'green' recycling was unclear with respect to standards and applicable rules. Therefore, a responsible ship owner who chooses upgraded facilities in China or Turkey could not be certain that the recycling was acknowledged and accepted as 'green' as defined in the Basel Convention and ILO Guidelines.

Today, a few shipping companies have health, safety and environment policies on recycling that will require 'green' facilities if and when one of their vessels must be scrapped. One company has used a European facility (BP Shipping) and another (P&O Nedlloyd, now Maersk Line) has carried out upgrade and training in Chinese yards for the scrapping of several of their vessels.

However, for the majority of shipping companies the added cost of 'green' recycling (in Asia approx. 15-25% of steel value) is equally important as the lack of certified recycling yards.

## OBJECTIVES

The overall objective of the study is to provide information and guidance to the European Commission on the issue of ship dismantling especially with respect to the environmental, technical and economic aspects of the problem and of the available options to contribute to a solution at a European level.

## INTERNATIONAL GOVERNANCE OF SHIP DISMANTLING

This stage provides an overview of the work and developments undertaken by the relevant IMO, Basel Convention and ILO bodies with regard to international governance on ship dismantling.

It contains a review of the draft IMO Convention on the Safe and Environmentally Sound Recycling of Ships being negotiated in the organisation's Marine Environment Protection Committee (MEPC) for adoption by a diplomatic conference in 2008/2009. The level of control and effectiveness in ensuring globally binding minimum standards for safe and environmentally sound recycling of ships is considered.

Outstanding key issues are flagged. With a view to compare the control level of the draft IMO Convention with that of the existing EU Waste Shipment Regulation, the stage is commenced by an outline of the main EU and international instruments relevant for ship dismantling.

In this context some recent examples of application of the Waste Shipment Regulation to ships for dismantling are provided.

## EXISTING EU AND INTERNATIONAL INSTRUMENTS RELEVANT FOR SHIP DISMANTLING

A broad range of EU legislation is applicable to end-of-life ships where it sets up binding environmental and safety requirements related to the dismantling activities themselves as well as the transfer of end-of-life ships for dismantling from and to the EU.

The Community waste legislation applies to the management of ships which have become waste. The Waste Framework Directive (Directive 2006/12/EC on waste ), sets out a number of requirements, including safeguards for environmental protection during recovery or disposal, planning and permitting requirements, record-keeping and periodic inspections.

This directive also lays down the definition of 'waste' in its Article 1(a) as "any substance or object in the categories set out in Annex I which the holder discards or intends or is required to discard". The shipment of end-of-life vessels within the EU or between its Member States and third countries are currently regulated by Regulation (EEC) No 259/93 on the supervision and control of shipments of waste within, into and out of the European Community (hereinafter the Waste Shipment Regulation).

The Waste Shipment Regulation determines which procedures to apply before a waste can be shipped within, into and out of the Community and is largely based on the United Nations' Basel Convention, as amended and the OECD Decision C (2001)107 on revision of decision C (92)39/final on the control of transboundary movements of waste destined for recovery operations.

The Basel Convention establishes a control procedure for the export and import of hazardous waste between the parties of the convention. The procedure is based on prior notification of the export and import and written consent from the concerned authorities before the import or export takes place.

To ensure that the State of import has the necessary information to make an informed decision whether to accept or reject an import of waste, the Basel Convention requires that certain information is provided to the State of import. This information is listed in Annex V,A to the Convention and includes detailed information of the waste in question, proposed methods of transportation and disposal, as well as evidence of a contract between the exporter and the disposer.

Confirmation is also required that the disposal contract specifies the environmentally sound management of the waste. The Basel Convention furthermore imposes an obligation (Article 4, para2) on each party not to allow export of hazardous waste or other waste and to prevent import of such waste "if it has reason to believe that the waste in question will not be managed in an environmentally sound manner".

The Convention thus seeks to prevent that waste be deposited in a State that has not consented to such a transaction or which does not have the capacity to deal with the waste in an environmentally sound manner. The Basel Convention also sets out specific steps to be taken (Article 8) when the notification procedure has been complied with, but the disposal cannot be completed in accordance with the terms of the disposal contract.

The Convention places the responsibility on the State of export, which must ensure that the waste is taken back by the exporter, if alternative arrangements cannot be made for environmentally sound disposal of the waste within 90 days from the time the importing State informs the State of export and the Basel Convention Secretariat of the inability to complete the disposal in compliance with the contract.

The Convention contains a list of hazardous wastes and a list of non-hazardous wastes. The parties to the convention may unilaterally designate other wastes as hazardous. It is an important feature of the Basel Convention control mechanism that it establishes a tracking system by which transboundary movements are monitored up to the issue of the final certificate of disposal.

It should be noted that disposal in the context of Basel Convention includes operations that lead to resource recovery, recycling, reclamation, direct reuse or alternative uses. In 1995, an amendment to the Basel Convention was adopted banning all exports of hazardous waste destined for recovery from OECD or EU members to countries to which the OECD decision does not apply. The ban amendment has not come into force yet, as not sufficient parties have ratified it. The EU has unilaterally implemented the ban (Council Decision 97/640/EC) meaning that the exports of hazardous waste from the EU to non-OECD countries are banned.

Articles 14 and 16 of the Regulation thus ban the export of hazardous waste to non-OECD countries. No exemptions are possible. Annex V to the regulation defines the wastes that fall under this export ban. End-of-life ships are not explicitly listed in Annex V. However, unless a ship has been built very recently it would in most cases contain hazardous substances listed in Annex V (e.g. asbestos under entries A 2050, 17 06 01 and 17 06 05 or PCBs under entries 16 01 09, 16 02 09 and 16 02 10) and would therefore be covered by the export ban.

The export and import of non-hazardous wastes are also covered by the Waste Shipment Regulation. For non-hazardous waste destined for recovery and contained in the "green list", it allows as a main rule that the destination country outside the OECD determines whether it wants to accept or ban the import of waste.

At international level, non-hazardous wastes for recovery are covered by an OECD Decision and are not within the scope of the Basel Convention. Annex II to the EU Waste Shipment Regulation lists all "green" non-hazardous wastes and implements in this regard the OECD Decision. Pursuant to this Annex II, ships are considered non-hazardous if they fall under the entry GC 030 'vessels and other floating structures for breaking up, properly emptied of any cargo and other materials arising from the operation of the vessel, which may have been classified as dangerous substance or waste'.

What the term 'properly emptied' means is not precisely defined in the OECD decision. It should be noted that export of non-hazardous waste destined for disposal in OECD countries other than EFTA countries which also are Parties to the Basel Convention is banned (Article 14). A recent Dutch judgement (of 21 February 2007) in the so-called *Otapan* case ruled that export of ships for dismantling which requires as a first step the removal and disposal of hazardous materials such as asbestos and PCB that cannot be recycled, constitutes an export for disposal.

The implication of the Dutch *Otapan* ruling, if applied in other cases, is that only ships which have been properly emptied can be dismantled in OECD countries other than EU and EFTA countries. The judgement is further analysed below.

To sum up, the Community legislation on waste, in particular the Waste Shipment Regulation, sets the requirements at EU level for management and shipments of end-of-life vessels. The export from Member States to non-OECD countries of end-of-life vessels considered as hazardous waste is banned according to the United Nations' Basel Convention as amended.

The ban is legally binding in the EU according to the Waste Shipment Regulation. It follows from Article 1(a) of Directive 2006/12 on waste and Article 2(a) of the Regulation that a ship becomes waste when it is discarded or is intended or required to be discarded – meaning from the moment that it is the intention of the owner not to use it as a ship anymore.

A ship destined for dismantling is thus to be considered a waste. Whether the end-of-life ship is considered as 'hazardous waste' depends on the vessel's contents of hazardous substances, and whether it has been 'properly emptied' of such substances.

## RECENT CASES OF APPLICATION OF THE WSR TO SHIPS FOR DISMANTLING

The subjective element of definition of waste (the holder's intention to discard it) sometimes has made it difficult to effectively apply and enforce the Waste Shipment Regulation and/or the Basel Convention to end-of-life ship.

One such example is the case of the Danish ferry Kong Frederik IX being sailed to India for dismantling. In February 2005 the Danish Ministry of Environment became aware that a 51 year old Danish ferry named Kong Frederik IX, later renamed Frederik and after that Ricky, was to be sold possibly for recycling in India. It also became known that the ship contained asbestos. When the ship was sold it was docked in Denmark, but registered in St. Vincent and the Grenadines.

The former owner of the ship, as well as the new owners, claimed that the ship was sold allegedly to be put in service in the Middle East as a cargo ship. The Danish authorities requested the involved parties to produce documentation on this.

As declarations and documentation for this activity were produced, the national competent authority classified the ship as a non-waste and the ship left the Danish port on 16 March 2005. However, after having left the Danish port and Danish waters the ship sailed directly for recycling in India. The ship reached India on 22 April 2005.

Denmark is of the opinion that the export of this ship should be regarded as an export of hazardous waste, and as such falls under the Basel Convention. Furthermore, Denmark believes that as the exporting State, it has a responsibility to re-import the ship.

India did not share this opinion. Denmark was not able to take action when the ship was anchored in the Danish port, since the new owner assured by a sworn statement that the ship was going to carry trailers as ro-ro cargo. By this sworn statement it was believed that the ship could not be classified as waste and hence could not be detained. The ship was subsequently dismantled in September 2005 in India. Whether the new owner decided to dismantle the ship while the ship was anchored in the Danish harbour (and as such provided the Danish authorities with false information), or whether this decision was taken in Danish waters or international waters, it has been argued that this case shows that we need an enforceable regime for ship dismantling.

The Waste Shipment Regulation has effectively been applied in a number of cases of export of ships for dismantling from EU Member States. This was the case of the Clemenceau, where the French air carrier was sent to India for recycling. It was made known to the Commission that the vessel had not been fully removed of its hazardous substances, especially asbestos.

The Commission requested additional information from the French government and considered launching infringement procedure against France rejecting France's argument that the Clemenceau is not covered by EU waste rules because it is a warship. Subsequently, the French government recalled the *Clemenceau* in order to prepare it for safe recycling following a ruling from France's highest administrative court ruled that a decision to send it to India for dismantling was illegal.

In the context of the Waste Shipment Regulation, the distinction between disposal and recycling has been a key issue in several cases. Most recently a Dutch Court (the Administrative Jurisdiction Division of the Council of State) on 21 February 2007 in the so-called Otapan case ruled that export of ships for dismantling which requires, as a first step, the removal and disposal of hazardous materials such as asbestos and PCB that cannot be recycled, constitutes an export for disposal.

The chemicals tanker *Otapan* left the Netherlands for Turkey at the end of July 2006. As the vessel had more asbestos on board than specified in the notification, Turkey refused to allow it to enter its territorial waters. The Dutch authorities (State Secretary for Housing, Spatial Planning and the Environment) then decided that the ship should return to the Netherlands. The proposed shipment was therefore not completed.

It was apparent from the documents that it was proposed to send the vessel back to Turkey once the quantity of asbestos aboard has been brought into line with the notification. The Dutch Court ruled that "the notification form states that the proposed shipment relates to a decommissioned tanker with asbestos-containing material on board.

This operation is described on the form as recovery as referred to in Category R4 of Annex IIB to the Directive. A plan for the scrapping of the ship was attached as an annex to the form. In the Division's opinion, it can be inferred from this that, contrary to what the defendant submitted at the hearing, the treatment of the waste specified in the notification is a process consisting of several stages.

The scrapping plan states that the first part of the operation is to remove the asbestos in the ship. It follows from the plan that the removal of the asbestos is necessary in order to enable the ship to be scrapped. The plan also emphasises that no other work on the ship, for example the removal of liquids and gases, should start until the asbestos has been removed. In view of the above the Division considers that the waste treatment process is wrongly classified on the notification form as a recovery operation as referred to in category R4 of Annex IIB to the Directive.

Nor is this altered by the fact that substances obtained from the dismantling of the ship will subsequently be transferred from the Simsekler yard to the metalworking industry for recycling, which could possibly be classified as a recovery operation.

In these circumstances it must be concluded that the defendant wrongly failed to object to the proposed shipment on the ground of an incorrect classification on the notification form". The implication of the Dutch *Otapan* ruling, if applied in other cases, is that only ships which have been properly emptied can be dismantled in OECD countries other than EU and EFTA countries, otherwise dismantling of EU ships can only take place within the EU.

## WORK BY IMO, BASEL AND ILO BODIES

The problem of ship dismantling has been discussed for many years within the international organisations involved: the International Maritime Organisation (IMO), the International Labour Organisation (ILO) and the Basel Convention, a treaty under the auspices of the United Nations Environment Programme (UNEP).

All three organisations have developed non-binding technical guidelines for ship recycling. A Joint ILO, IMO and Basel Convention Working Group on Ship Scrapping has been established.

It has held two meetings in 2005. At the meetings the Joint Working Group has discussed a co-ordinated approach to the issue in order to avoid duplication of work and overlapping of roles, responsibilities and competencies between the three organizations.

The Group has undertaken a comparison of the technical guidelines of the three organisations. It has encouraged collaboration in their technical cooperation activities and has identified activities that could promote implementation of the guidelines.

The Joint Working Group has furthermore addressed how a coordinated approach to all relevant aspects of ship scrapping may best be undertaken. In this context, the concept such as environmentally sound management and prior informed consent has been discussed and has, as further elaborated below, discussed an initial comparison between the Basel notification procedure and the proposed reporting system under the draft IMO Convention on the Safe and Environmentally Sound Recycling of Ships being negotiated in the organisation's Marine Environment Protection Committee (MEPC).

## REVIEW OF THE DRAFT IMO CONVENTION

Since 2005, the IMO has been working towards a binding international regime for clean ship dismantling. A draft Convention on the Safe and Environmentally Sound Recycling of Ships is being negotiated in the organisation's Marine Environment Protection Committee (MEPC) for adoption by a diplomatic conference in 2008/2009, to enter into force some years later.

It aims at a "cradle to grave" approach to ship recycling, with a view to avoiding hazardous materials in new ships and removing them from existing ships during their period of operation. It will set out specific requirements for the ship recycling facilities including authorisation of the facilities as well as a reporting system.

Technical details, which also cover the environmentally sound management of ship recycling facilities, will be set out in recommended guidelines supplementing the Convention. A large number of provisions are still to be agreed to.

There are thus a rather large number of square brackets. The structure of the draft Convention is linked to the procedures for amending the Convention as defined in Article 18, which specify explicit acceptance by two thirds of the Parties to the Convention for amending Articles and a tacit procedure for amending the regulations contained in the Annex, whereby amendments would be deemed accepted unless one third of the Parties to the Convention notify their objection.

The following guidelines are foreseen to be developed:

- General
  - Guidelines for communication of information.
- Guidelines for ships
  - Guidelines for the development of Inventory of hazardous materials.
  - Guidelines for the submission of a proposal to control hazardous materials.
  - Guidelines for surveys and certification.
  - Guidelines for inspection of ships.
  - Guidelines for establishing gas-free-for-hot-work conditions.
- Guidelines for ship recycling facilities
  - Guidelines for authorization of ship recycling facilities.
  - Guidelines for safe and environmentally sound ship recycling.
  - Guidelines for the development of S hip Recycling Plan.

In order to further develop the draft Convention and its guidelines, MEPC 55 in October 2006 agreed to convene an intersessional correspondence group and also an intersessional working group. The intersessional Correspondence Group, under the coordination of Norway would report on the outcome of its work to the intersessional Working Group scheduled to meet in May 2007 in the UK which would provide a report on the draft convention and related guidelines for further consideration by MEPC 56 (July 2007). Following MEPC 56, the twenty-fifth regular session of the Assembly (3-14 December 2007) will decide on the holding of the Diplomatic Conference.

MEPC 57 in March 2008 is scheduled to establish a Working Group and possibly to finalise work on the draft Convention. The one week Diplomatic Conference may be held sometime between February and April 2009. Controversial issues, which will be discussed by IMO working groups in 2007, include the question as to whether reference should be made to rules and standards outside the framework of IMO, the environmental baseline standard for ship recycling facilities, future reporting requirements (in particular whether state-to-state notification should be necessary as under the Basel Convention) and the type of compliance mechanism to help with implementation of the Convention.

In the context of the Basel Convention, the key issue under discussion is whether the proposed Ship Recycling Convention will ensure an equivalent level of control and enforcement as under the Basel Convention. The 8 Conference of the Parties on 1 December 2006 reiterated this demand and stated, among other things, that the future legal instrument should generate conditions of ship dismantling that protect workers and the environment from the adverse impacts of hazardous wastes and unsafe working practices.

The Council in its conclusions of 21 November 2006 "emphasised that ship that are waste will continue to fall under the EU Waste Shipment Regulation as long as there is no internationally binding and effective legal instruments, which while addressing any special characteristics of ships, guarantee an equivalent level of control and enforceability as established under the Basel Convention".

The Council also underlined that the future IMO Ship Recycling Convention should "generate real changes in the conditions under which "end-of-life" ships are dismantled, so as to protect workers and the environment from the adverse impacts of hazardous waste and dangerous working practices". In the following, the level of control and effectiveness in ensuring globally binding minimum standards for safe and environmentally sound recycling of ships is considered.

Focus is placed on the draft provisions (as of January 2007) which have been agreed to and key outstanding issues are flagged. According to the current draft, the Convention - in line with other IMO instruments - will not apply to smaller vessels of less than 500 gross tons, nor to warships, naval auxiliary or other vessels which are state-owned or - operated and used only on non-commercial government service. It is provided, however, that "each Party shall ensure, by the adoption of appropriate measures, that such ships act in a manner consistent with this Convention, so far as is reasonable and practicable." The draft Convention sets out a number of mandatory requirements for ships with a view to avoid hazardous materials in new ships and removing them from existing ships during their period of operation.

Each new ship shall thus have on board an Inventory of Hazardous Materials (Regulation 7). The inventory shall be approved by the Administration taking into account the guidelines developed by IMO.

The inventory shall be specific to each ship and shall at least:

- identify hazardous materials, their location and approximate quantities;
- consist of Part I: materials contained in ship's structure and equipment;
- clarify that the ship does not contain any materials which are prohibited under regulation 4.

The provisions for inventory for hazardous materials for existing ship have not yet been agreed to. There are currently two alternatives. The objective of the inventory of hazardous materials is to provide specific information on the location and quantities of potentially hazardous materials on board each ship so as to protect the health and safety of workers and also to prevent environmental pollution at the recycling facilities.

Inventories will consist of three Parts:

- Part I: Materials contained in structures and equipment of the ship;
- Part II: Operationally generated wastes: and
- Part III: Stores.

The inventory is supplemented by a survey and certificate regime (Article 5 and Regulations 11 - 15). Presently, the draft Convention makes provisions for the following surveys:

- one initial survey; for new ships an initial survey would take place before the ship enters service, while for existing ships the survey would take place before the inventory is issued (existing ships are defined as ships built before the Convention enters into force);
- periodical surveys, on the basis of a five year cycle, verifying continuing compliance with the Convention and ensuring that any relevant alterations are reflected on the inventory;
- additional surveys, which the ship owner may request to take place at his option after alterations to the structure, machinery or equipment of the ship; and
- one final survey prior to the ship being taken out of service, so that all three Parts of the inventory are completed and checked, the Recycling Plan is examined and checked against the inventory, and the Ready for Recycling Certificate is issued.

The most important requirements for recycling facilities are

- mandatory development of ship recycling plan. The plan shall be developed by the recycling yard to specify the manner each ship will be recycled, depending on its particulars and its inventory. It shall take into account the non binding guidelines developed by IMO;
- the introduction of the International Ready for Recycling Certificate; to be issued to the ship following its final survey verifying ship related statements on the Recycling Plan;
- the authorization of recycling facilities by their States, in accordance with the requirements of the Convention; and
- a Statement of Completion of Ship Recycling, issued by the recycling facility and reporting completion of recycling of an individual ship to the recycling State authorities and to the flag Administration.

The issue of pre-cleaning has been a divisive issue. Under the draft Convention, prior removal of hazardous materials may not be required if the recycling facility chosen is fully authorised to manage the type or amount of hazardous materials contained in the ship.

The draft Convention seeks to match the individual ship's requirements and the capabilities of the recycling yard by means of a Recycling Plan and an International Ready for Recycling Certificate. However, there is still a need to clarify that, in case the final recycling facility is not capable of managing certain hazardous materials, the owner of an end-of-life ship must select another yard or remove the materials for safe disposal or recovery beforehand.

As indicated above the future reporting requirements (in particular whether state-to-state notification should be necessary as under the Basel Convention) and the type of compliance mechanism to help with implementation of the Convention are some of the controversial issues to be discussed at the working ground meetings in 2007 and key outstanding issues assessing the level of control and enforcement.

While the mandatory system for inventory of hazardous materials and the associated survey and certificate regime sets out comprehensive requirements which may effectively be applied to all ships and thus generate change on the ground for the conditions under which end-of-life ships are dismantled, the current draft does not address in operational terms how new survey and certificate will be issued if the ship changes owner and/or flag. This is particular relevant if the ship changes owner just prior to the recycling.

A number of provisions are relevant for control of the recycling facilities (Articles 4 (2), 6, 7, 8(1), 9(4), 10 (1), 12 (2) and (6), 13bis, 17 and 18 (2). Several of these provisions are of a generic nature.

Furthermore unlike a number of other IMO legally binding instruments, such as MARPOL on Port Reception Facilities, the present draft IMO Ship Recycling Convention does not provide for dissemination of information on the web on IMO which often is seen as an effective mean to facilitate compliance and enforcement.

Finally while the IMO Convention may eventually provide for globally binding minimum standards for safe and environmentally sound recycling of ships, it should be noted that IMO conventions on average take six years to enter into force.

The Convention would therefore most likely enter into force in 2013, at the earliest. The peak for phasing-out single hull tankers is expected around 2010.

## UPDATED STATUS AND PROJECTIONS FOR EUROPEAN END-OF-LIFE SHIPS UNTIL 2020

The number and tonnage (LDT) to be scrapped until 2020 is assessed below and a status is provided for the historical developments with a special emphasis on the years 2004-2006. For analytical purposes, the scrapped vessels have been split into eight *type segments.*

The segmentation is based on a categorisation of the "sub types" used in Clarkson's databases consistent with the methodology used in the COWI/ TREN study (2004). The assessment of future decommissioning volumes also includes two additional types *Fishing vessels* and *Warships* and the assessment is therefore made for the following 10 'vessel types':

- (Single hull) Oil tankers
- Other tankers
- Bulk
- Container
- Gas
- Passenger/ro-ro/vehicle (including ferries)
- Other cargo vessels
- Non-cargo vessels
- Fishing vessels
- Warships

No distinction is made between the hull types of oil tankers.

## HISTORICAL DECOMMISSIONING VOLUMES

This stage describes the recorded scrapping activity from 1994-2006, with a special emphasis on the years 2004-2006.

### Total volumes 1994 – 2006

The estimated level of historical scrapping are summarised in the figure below.

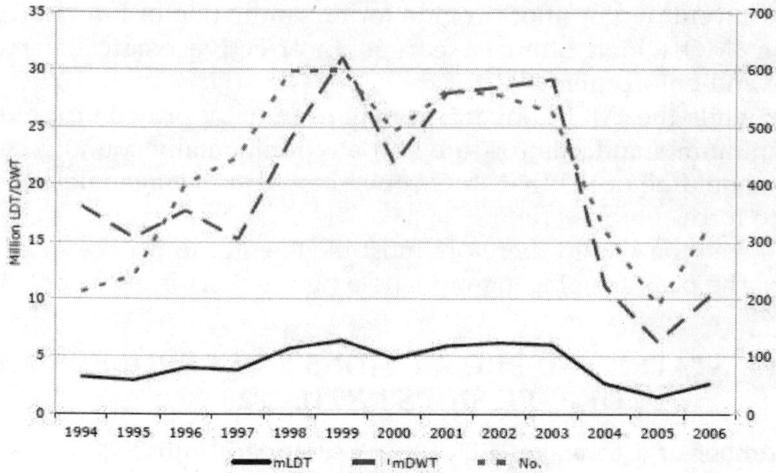


**Fig. Total historical ship scrapping volumes, all types**
**Million LDT, Million DWT and number of vessels**

From 1994-2006, approximately 5,600 ships have been demolished worldwide. There have been considerable variations in the activity level over the years.

The ship scrapping activity peaked in 1999 with 600 ships being scrapped representing approximately 6.4 million LDT, while the scrapping activity in 2005 reached an "all time low" of only approximately 1.5 million LDT being scrapped.

## Volumes by ship type

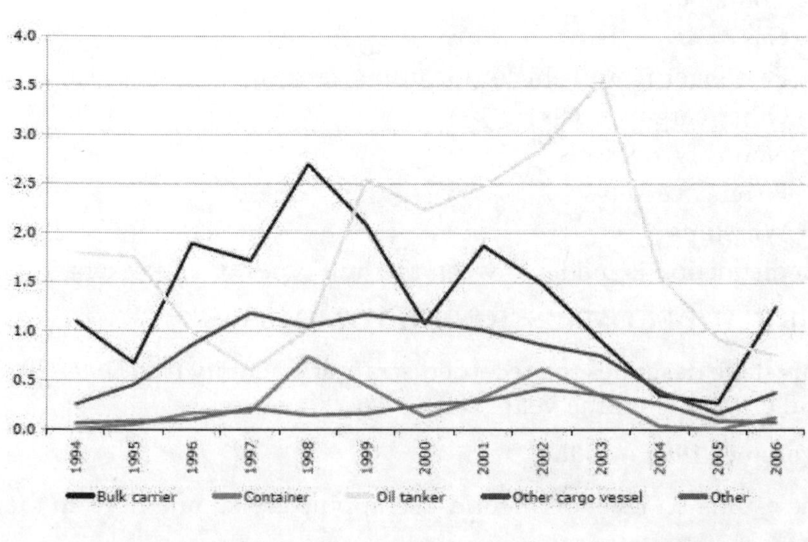

The historical scrapping volumes by ship type are shown in table below. The volumes scrapped declined considerably from 2003 to 2004 and 2005 for all major ship types due to historically strong freight markets.

It appears - on the basis of data for the first half of 2006 - that scrapping volumes are increasing due to relatively strong drops in the freight rates. It is seen in the above figure that the decommissioning volume has dropped for all ship types, but in particular the oil tankers are trading beyond their expected life time.

There are no EU records of the destination of warships, but usually the vessels are scrapped nationally or previously used for target practise, although this is not common today. The end-of-life destination for some larger vessels in recent years:

- In 2003 the frigate *HMS Scylla* was used as an artificial reef off Plymouth (UK).
- The German destroyer *Rommel* was scrapped in Turkey in 2004.
- The destroyer *Warszawa* and several other naval vessels were scrapped in Poland in 2004-2006.
- The training vessel *Lucifer II* is currently tendered for dismantling in Cherbourg, France, where scrapping of submarines also takes place.

Historically, in many European navies vessels are often sold for continued use in friendly nation's navies, but for obsolete vessels it was not uncommon to sell for the scrap value to Asian ship breakers.

## METHODOLOGY FOR ANALYSIS OF CURRENT FLEET AND FUTURE DECOMMISSIONING VOLUMES

### Approach and assumptions

A number of factors influence the future supply of vessels to the ship scrapping industry including the size of the existing fleet, the future earning potential, the cost of keeping the ship in operation, the age profile of existing fleet, the size of the current fleet, regulatory issues, budgets etc. The importance of the factors differs in vessel types and several approaches are used for the estimate of the current fleet and the projections.

- Vessel type 1. To facilitate comparison the single hull oil tankers are assumed to be phased out according to the same criteria as in the COWI/TREN study (2004).
- Vessel type 2-9: These vessels are operating on commercial conditions and are not subject to phase-out regulation. They are phased out according to the type's historical trading life expectancy, except for fishing vessels, which are estimated from other non-cargo vessels.
- Vessel type 10, Warships: EU warships are obviously not operating

on commercial conditions, but on the basis of the costs of keeping the vessel operating, the need for warships, political arguments, available budget etc. Regarding the future volumes of scrapping of warships it is simply assumed that all warships aged 25 or more will be scrapped during the next 14 years (2007-2020).

***Size limits***

The analysis focuses on vessels of 2,000 DWT and above for vessel type 1-8, fishing vessels of 500 GT and above and warships of 150 LDT and above. For the latter two types LDT is typically not given and conversion factors to derive LDT apply: LDT equals 54% of the GT for fishing vessels and 46% of the fully loaded weight for warships.

***Country grouping***

The projections of the future volumes of scrapping are made by flag state and country of ownership on the basis of the following definition:

- *EU countries*: The 27 EU Member States.
- *Non-EU countries*: All other countries (including accession countries).

This implies that projections are made for 4 categories:

- EU flag/EU owned
- Non EU-flag/EU owned
- EU flag/Non-EU owned
- Non EU-flag/Non-EU owned

The analysis covers the global fleet of vessel types 1-5 and 7-8, but only fishing (vessel type 9) and naval vessels (vessel type 10) which are flying the flag of an EU Member State and passenger vessels (vessel type 6) sailing to and/or from an EU country.

The country of ownership for vessel types 1-5 and 7-8 is identified based on the information provided in the Clarkson database, while the projections for fishing vessels (vessel type 9) are based on the Community Fleet Register. It has not been possible to determine the country of ownership directly for passenger ships (vessel type 6).

Instead it is assumed by looking at the vessels on each route in EU, that half the vessels are owned by the country of origin and half by the country of destination.

The information on flag state and route specific information is based on data from Shippax 'Statistics and Outlook 06'. Data on European warships are obtained from Jane's Fighting Ships, Yearbook 2006-2007.

## THE CURRENT FLEET

The composition of the current fleet of vessel types 1-10 is presented in this section. This information is used for estimating the future volumes of scrapping.

## Vessel type 1 - Single hull oil tankers

### *All owner and flag state countries*

The total fleet of single hull oil tankers represents 19.8 million LDT, of which 3.3 million LDT are category 1 tankers. The equivalent numbers for category 2 and 3 tankers are 14.3 million LDT and 2.2 million LDT, respectively. Note that no distinction is made between single hull oil tankers equipped with double bottom (DB) and single hull oil tankers equipped with double sides (DS) as these are treated equal in the regulation. A simple adjustment is made to counteract this artefact. Here it is simply assumed that these vessels are scrapped over the coming 5 years (ref. section 3.5). The age profile of the existing fleet of single hull oil tankers is presented in the figure below.

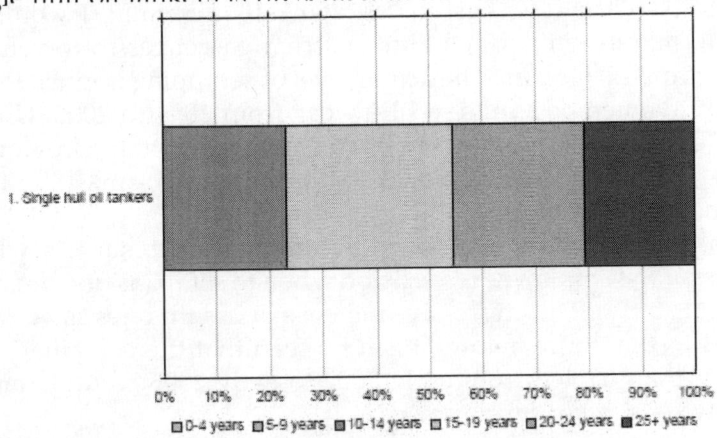


**Fig.** Age profile of current fleet of single hulled tankers—All owner and flag state countries, Vessel type 1 (single hull tankers) share of LDT. Due to the MARPOL recommendations no new single hilled tankers have entered the market for 10 years.

### *Current fleet by owner/flag state*

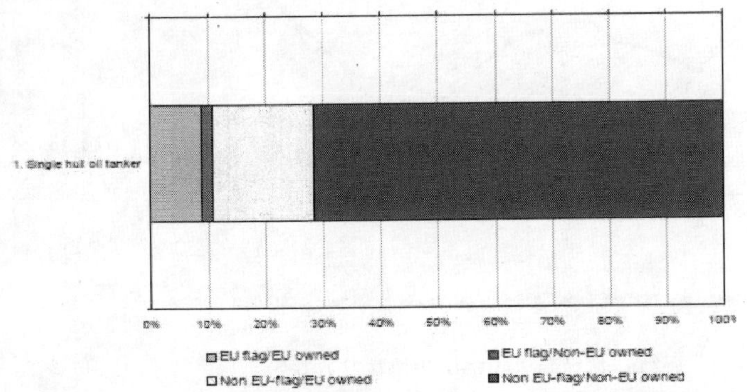


**Fig.** The current fleet of single hulled tankers—By owner/flag state. Vessel type 1 (single hull tankers) in million LDT

The majority of the single hull tankers are owned by residents of non-EU countries, and only a minor share of the fleet is flying an EU-27 flag.

## FUTURE SCRAPPING VOLUMES

### Earlier projections

The COWI/TREN study included simple projections for the future scrapping activity for 2003-2015. The projections were based on a simple assessment of the age profile of the fleet (for all other vessel types than oil tankers) and the historically observed life time expectancy. For oil tankers the consequences of different phase-out schemes were assessed, i.e. the IMO MARPOL 13G and EC 1726/2003/Revised MARPOL Annex 1.

For the years 2004-2006 it was estimated that 7-10 million LDT would be scrapped per year, mainly consisting of bulk carrier, passenger/ro-ro/vehicle, other cargo vessels and oil tankers. The actual level of scrapping is much lower approximating an average of 2 million LDT/year from 2004 to 2006. Unless the average life expectancy remain several years higher than previously observed, this has obviously created a back log of until now some 15 million LDT to be scrapped in the future.

The main reason for this is as mentioned the strong freight markets, which is the main driver for the ship owners' decision when to scrap (as documented in the COWI/TREN study). The impressive increase in freight rates since 2002/2003 is illustrated in the figure below. It can be seen that the container time charter rates more than tripled from early 2002 to early 2005 and the tanker segment has shown a similar development.

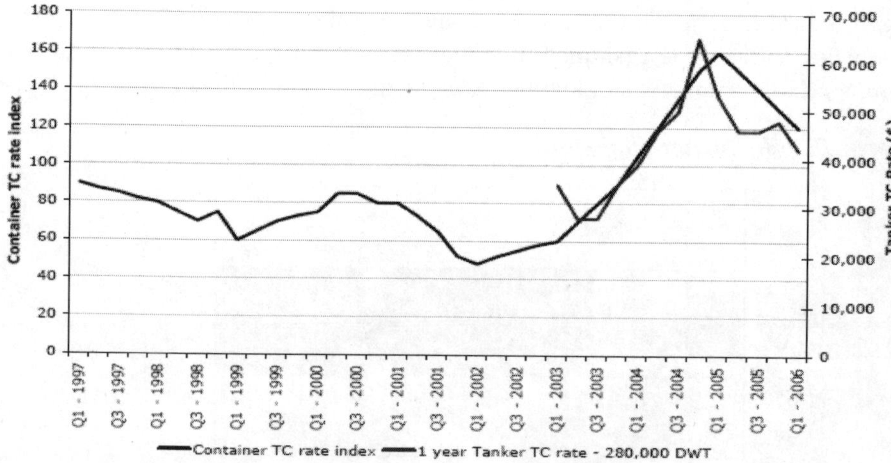


**Fig.** Freight (time charter) rate.

It is emphasised that the present models of phase out do not include forecasts of freight rates and projects scrapping based on historical phase out patterns. Therefore, the recent years' high freight rates leading to continued trading of vessels previously considered at their end-of-life were not predicted and likewise a sudden drop in freight rates with subsequent massive scrapping will not be predicted.

**Total fleet**

The estimates of the future total volumes of demolition are shown in the Figure below. The volumes of scrapping are estimated to fluctuate between around 5-10 million LDT/year during 2007-2020, except for 2010 where a large number of single hull tankers are phased-out.

In total, it is estimated that 105 million LDT will be scrapped from 2007-2020. The future volumes of scrap are estimated to start at a relatively high level for the next few years and then show a declining trend until 2016. From 2017 and onwards, the volume of scrapping is estimated to show an increasing trend.

Not surprisingly bulkers, single hull oil tankers, container vessels and other cargo vessels account for the vast majority of future scrapping. The annual decommissioning volumes of fishing vessels are projected to decrease until 2010 where after the annual volumes remains stable between 40,000 and 45,000 LDT. As mentioned previously it is virtually impossible to forecast the future level of scrapping of warships.

As an example the UK naval authorities will decommission approximately 30 naval vessels before 2013, but the majority of these are expected to be sold to foreign governments or to other uses that recycling (UK Defra 2006). The legacy ships that are expected to be recycled comprise five out of the thirty vessels.

Two of the vessels are up to 20,000 tons. In general, it seems fair to assume that warships which today are 25 years or older will be scrapped during the next 14 years (2007-2020). The fleet of European warships aged 25 or older represents 396,000 LDT.

Hence annual scrapping of warships can be expected to be in the region of 28,000 LDT. In lack of suitable information it is assumed that the sale of navy vessels, which is not subtracted here, will balance the added vessels belonging to other government agencies.

In the French report on demolition in European military fleets it is estimated that 300,000 LDT may be dismantled over the coming 10 years (MIDN 2007).

The sum of the projected annual scrapping volume for European warships and larger fishing vessels is approx. 70,000 LDT and compared to the merchant fleet it only account for a very small fraction (approx. 1%) of total scrapping.

As actual volumes are sensitive to the developments in the freight market, these numbers should only be seen as indicative for the level of scrapping.

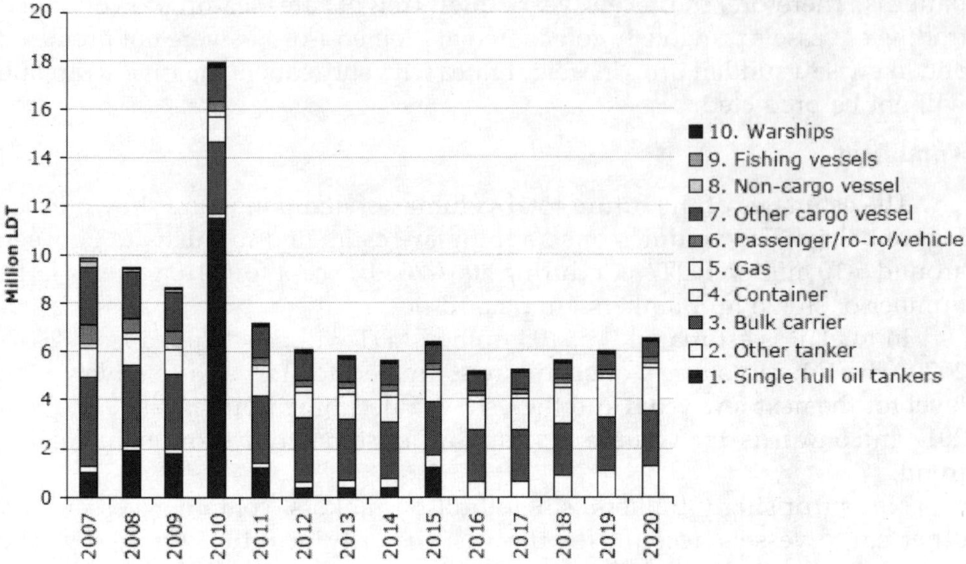


**Fig.** Future volumes of demolition by vessel type and year of scrap Vessel type 1-10 (Million LDT)

## By owner/flag state

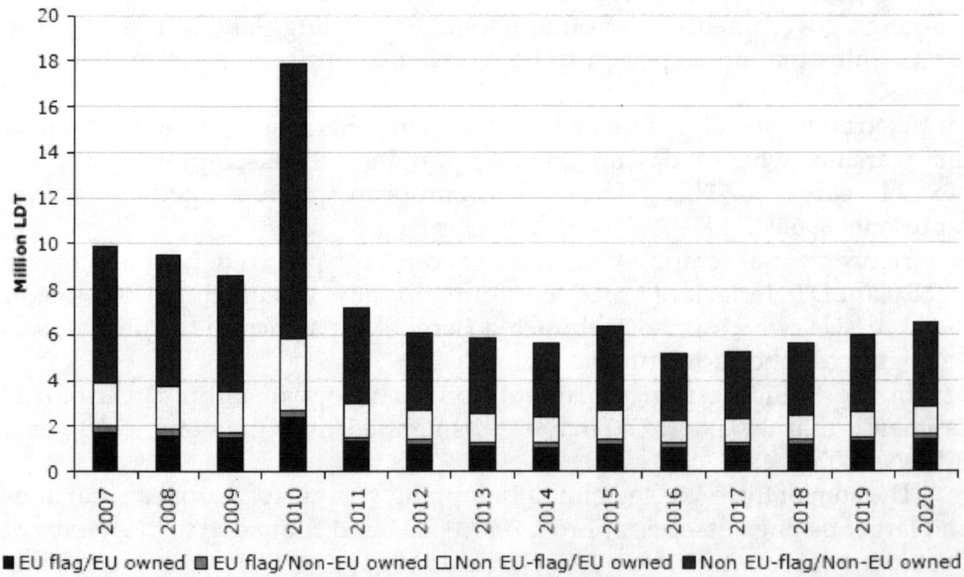


**Fig.** Future volumes of demolition by owner/flag state and year of scrap—Vessel type 1-10 (Million LDT)

Forecasts by country of ownership and flag state are shown in Figure below. Vessels owned by non-EU countries and flying the flag of non-EU countries account for the largest share of future scrapping - in the region of 3-6 million LDT/year, except for the peak in 2010 where a large number of single hull oil tankers are phased-out.

Scrapping of vessels, which today are both EU flagged and EU owned, will amount to 1.1-1.8 million LDT/year, except for 2010. EU owned vessels flying the flag of a non-EU country accounts for in the region of 1.0-1.9 million LDT/year. Finally, EU flagged vessels which are owned by non-EU countries, will only account for scrapping of 0.2-0.3 million LDT/year.

The share of vessels operating under a European Union member state flag is 22% measured as LDT. The vessels operated or owned by companies in the EU accounts for 40% and the combination accounts for 43% of the global LDT.

## UPDATED STATUS OF CURRENTLY AVAILABLE AND PROJECTED SHIP RECYCLING AND PRE-CLEANING CAPACITY

The stage examines the existing capacity for recycling of ships under safe and environmentally sound management conditions in the countries mentioned below.

- the 27 EU member states,
- the EU candidate countries Turkey and Croatia,
- the rest of the OECD and
- the non-OECD shipbreaking nations in Asia (Pakistan, India, Bangladesh and China).

The technical options for increasing the capacity of green recycling are discussed, and also addressed is the supply side to recycling in the EU and elsewhere, i.e. the vessels and the barriers for increasing their availability to recycling yards. It is not attempted in this report to define the terminology regarding the reuse of ships.

The major institutions and organisations have favoured their own terms: scrapping, dismantling, breaking, decommissioning and recycling. When the different terms are used in this report they are not implying a particular methodology, but rather the origin of the information treated. Since a common effort is carried out on an IMO Convention on Ship Recycling, this term is used in general.

To emphasise the need for improved or internationally recognised conditions, the phrases "safe and environmentally sound" or "green recycling" are favoured. The European Commission in 2001 published a study on Technological and Economic Feasibility Study of Ship Scrapping in Europe.

This study concluded that establishing dismantling capacity in Europe was not economically viable on market conditions. In the COWI/TREN report of 2004 a limited number of European facilities listed at the time were engaged in shipbreaking or under establishment. Their business plans were typically based on government vessels in combination with offshore oilrigs decommissioning rather than the merchant fleet. The present update shows that the majority of the planned facilities did not mature into active ship recycling facilities or they are still in the process of obtaining the necessary permits and licences from local authorities.

However, the continued public pressure for avoiding the scrapping practises of Pakistan, India and Bangladesh, the phase out of single hulled tankers, and presumably the increased price on recycled steel have initiated more plans for European ship recycling facilities. It is clear that European ship recycling is still limited since practically all merchant vessels of size and scrap value are broken in Asia.

It has also been stated in previous cost analysis that under the current legislative regime the economic feasibility of recycling yards based on merchant vessels are unrealistic. The mere lack of facilities and limited success in establishing new ones in the EU is a statement of this. It must be emphasised that a yard's capability to perform "green" recycling is very difficult to establish as there is no internationally recognised system for certifying recycling accordingly.

In the contacts to industry in the present study the capacity is examined by the documentation made available by the yards or interviews with selected yards. The references to a yard's certification (e.g. ISO, OSHA) and of approved, certified or similarly recognised companies and facilities for disposal of hazardous waste are taken as indication that "green" recycling is possible.

## FACILITIES FOR GREEN RECYCLING IN THE EU

In the wake of the unsuccessful attempts to have the aircraft carrier *Clemenceau* scrapped in India or Turkey, the French government initiated an extensive survey of European and overseas facilities with a potential for dismantling a vessel of this size under appropriate concern for safety, health and environmental issues.

The French Inter-ministerial Working Group on Ship Dismantling presented their draft findings during an EMSA workshop in September 2006 with the conclusion regarding Europe that: *There are some recycling capabilities of vessels in Europe (Belgium, the Netherlands, Italy, Lithuania and potentially Great-Britain) as well as in the others OECD countries (Turkey and eventually Norway).* It was clear that none of the sites visited by the French group were immediately ready to commence dismantling of a vessel the size class of *Clemenceau*, but with due preparation the technical feasibility was achievable at several sites across Europe (Lebacq 2006). This may also include sites in Le Havre and in Cherbourg, France, not previously engaged in recycling of large ships

(Assemblée Nationale 2007). The active European (EU-27) ship recyclers with existing capacity for vessels up to 150 m includes:

- Van Heyghen Recycling S.A. in Gent, Belgium, with a capacity of up to 60,000 LDT/year and a length of 265 m;
- Scheepssloperij Nederland B.V. in s'Gravendeel, the Netherlands, with a capacity of 30,000 LDT/year;
- Gdynia shipyard, in Gdynia, Poland has recently dismantled several naval vessels, including the destroyer "Warszawa". Capacity unknown;
- Fornaes Aps in Grenaa and Smedegaarden in Esbjerg, both Denmark, have capacities of 30,000 and 20,000 LDT/year;
- Desguaces de la Arena (Soto del Barco) in Spain can dismantle vessels up to 220 m length, but draught is presently limited;
- Simont S.p.a. in the Port of Naples, Italy has a capacity of 80,000 LDT/year and has previously recycled large naval vessels. The company is presently most active in industrial demolition.

The present situation in the Campania region in Italy, under which jurisdiction Naples is placed, is that there are problems with waste management capacity, so that imports of waste from other regions are generally not allowed.

BP Shipping states in their annual report that three facilities are used and inspected by them in Spain. The annual capacity of these facilities has not been established, but the capability to perform green recycling has been demonstrated by the company. It can be derived from the experience of the recent French study, the UK strategy, and the previous findings of the Danish study on OECD capacity that the level of implementation of EHS standards does vary in the visited yards, also within the EU.

However, it is estimated that all yards in the course of one year would be able to comply with the requirements of the guideline of the Basel Convention (UNEP 2004).

## Planned or dormant capacity for green recycling

### *Smaller facilities in EU*

A number of facilities exist in most sea locked European countries for recycling of smaller vessels such as fishing vessels, barges, small naval vessels etc. E.g. in Poland 150 fishing vessels were recently scrapped in various locations along the Baltic coast (GdaD□sk, Gdynia, WB□adysB□awowo, KoB□obrzeg, Dziwnów), in Latvia and Lithuania several facilities recycle smaller vessels, and on the northern coast of Spain a number of pier breaking yards exist. In Bulgaria two facilities operate at Varna and Bourgas; in Greece two facilities (Bacopoulos and Savvas Pireus) dismantles smaller ferries, fishing and naval vessels. The upgrade and expansion of these smaller facilities to green recycling of larger vessels will depend on the feasibility for physical expansion in the individual settings of each company, but also on the methodology used. It will often be relatively simple to move a pier breaking activity to another quay, whereas a facility using a dry dock or a synchrolift will be costly to expand.

### *Ship dismantling facilities under establishment*

Two facilities in the United Kingdom are expected to be very close to come into operation in ship dismantling: The Able UK in Hartlepool and Harland & Wolf in Belfast. Both these sites have applied for ship dismantling licences to expand activities on existing sites and both will have the opportunity to accept larger vessels.

The shipyard of Harland & Wolf possesses the largest dry docks in Europe which can accommodate vessel of any known size. Able UK, who is ready to dismantle vessels from the US Ghost Fleet, expected the necessary permits in the near future at the time of writing. Bartin Recycling in France, known from its activity in recycling airplanes, and the company Europlasma have formed a consortium to recycle ships, in particular the aircraft carrier Clemenceau. In the Baltic countries Latvia and Lithuania several companies are engaged in small vessel recycling or ship repair and maintenance. Several of the yards are engaged in feasibility studies, expanding existing or establishing new recycling facilities for larger vessels:

- Latvia: Liepajas Metalurgs, Liepaja Kuusakoski, Riga
- Lithuania Klapeidos Laivu Remontas, Klaïpeda Kuusakoski, Klaïpeda Subare, Klaïpeda Undoris JSC, Klaipeda

### *Facilities for decommissioning of offshore structures in the EU*

It is still the case that the facilities established for decommissioning of offshore oilrigs and storage platforms often can be used for ship recycling. The facilities potentially available are:

- Shetland Decommissioning Company Ltd. in Greenhead Base, Lerwick Harbour, Shetland.
- KBR Caledonia, Nigg yard, Tain.

Using these facilities for recycling of ships will depend on the actual location and layout of the facility. It is clear that the companies involved in decommissioning offshore platforms do not see a profitable market in ship recycling under the current market conditions.

## Projection for EU-27 facilities

### *Government and fishing vessels – the steady supply*

It is evident that smaller yards exist in most EU countries facilitating the dismantling of inland and coastal vessels not suited for open sea travel to the major recycling areas of the World. The seven listed companies ready to scrap vessels up to 150 m and for one yard up to 220 m are sufficient for the vast majority of the vessels on the existing European recycling market, and to these can be added a few yards now ready to enter the recycling market with even larger facilities.

Even without the latter the capacity is more than 150,000 LDT. Obviously, some of this capacity is already utilized for small vessels, but judging from the example from Poland, the capacity for smaller vessels is readily developed in other ports if a yard decides to take on the larger vessels. Fishing vessels supply are projected to decrease from up to 70,000 LDT before 2010 where after the annual volumes stabilises between 40,000 and 45,000 LDT. The life time expectancy of warships, at least in peace time, is generally quite long (30 – 50 years).

It has been assumed that warships, which today are 25 years or older, will be scrapped during the next 14 years (2007-2020). The fleet of European warships aged 25 or older represents 396,000 LDT and 321 vessels. Hence, the annual scrapping of warships for the EU can be expected to be in the region of 28,000 LDT and 23 vessels.

Given the total affirmative supply on EU basis of less than 100,000 LDT/ year from naval and fishing vessels and the scarcity of very large government owned vessels, it is technically and logistically possible to accommodate the need for recycling of vessels in the EU in the projected future of ten years. In interviewing several yards it was also clear that even though they may currently experience too little draught in their facility to accommodate vessels of >7m, the yards considered dismantling topsides and freeboard outside the yards in port areas.

As mentioned earlier not all yards appear to fully comply with the requirements of the guideline of the Basel Convention, but is it estimated that all would be able to in the course of one year. In conclusion, for the limited supply of vessels from government and from the fishing industry, no technical constraints are foreseen for the EU capacity in ship recycling.

However, the scrapping of government vessels may well be seen by local yards, the public and local/national authorities as an activity well suited for job creation programmes in former ship building communities. It is therefore uncertain whether an EU wide market for decommissioned government vessels will emerge. This possible supply constraint may seriously limit the viability of green recycling yards in the EU. To counteract a yard would presumably seek to acquire vessels from the merchant fleet.

***The merchant fleet***

The projected volume for demolition for all the vessel types was given as (excl. the peak year 2010):

- 1.1-1.8 million LDT/year for EU flagged and EU owned (including fishing vessels and warships)
- 1.0-1.9 million LDT/year for EU owned vessels flying the flag of a non-EU country;
- 0.2-0.3 million LDT/year for EU flagged vessels which are owned by non-EU countries.

The vessels potentially up for recycling obviously comprise the part of the fleet close to their end-of-life, but it must be recalled that commercial vessels under EU flags are often sold to companies registered in non-EU countries a few or several years before their end-of-life. At present, there are limited possibilities for imposing a legal obligation on EU owners flying non EU flags to supply obsolete vessels to recycling yards of a certain standard acknowledged or certified internationally.

Measured on LDT approx. 22% of all vessels operate under an EU member state flag and some 40% is registered as owned by an EU based company. Virtually none of these vessels are scrapped in Europe, where the recycled vessels only accounted for approx. 1% of the global decommissioning volume. Assuming that the projection of vessels under European flag (1.3-2.1 million LDT) will also be scrapped in the EU, then the size profile of the vessels must be considered.

Basically, recycling of a ship requires a space large enough for the structure, mechanical equipment in various levels of sophistication, and workers trained in safe recycling. The two latter are easy to obtain, whereas the former may be more challenging depending on the standards required. Advantages of establishing ship recycling facilities in the UK are given in the Defra report (Defra 2007), but they will apply on an EU scale as well:

- The possibility of investment in safer, more efficient mechanisation of many of the processes undertaken manually in developing countries;
- Removal of hazardous materials by staff with appropriate expertise, using safe and environmentally sound techniques and equipment. The UK has access to such staff and equipment;
- A highly developed domestic and EU legal and regulatory infrastructure and effective enforcement of standards; and,
- A tradition of market-led innovation and development providing good opportunities for British industry to benefit from the worldwide demands for environmentally sound ship recycling facilities.

Any requirements of technical modification of existing yards and pier breaking activities as a result of international agreements, e.g. a coming IMO Convention, should be readily achievable. Today compliance to safety, health and environmental standards in ship recycling typically of smaller vessels are achieved by way of member state licensing and monitoring. There is no reason to believe that larger facilities employing the known methodology should exceed the capacity of authorities' inspectors and regulators.

***Dry dock recommendation***

If it is considered mandatory to use a dry dock for green recycling, there will be a constraint to consider, in particular for the very largest vessels. The large dry docks in the EU (e.g. Gdynia, Poland; Constantza and Mangalia, Romania) and the very large docks (e.g. Belfast, United Kingdom; Hoboken, Belgium; Cadiz, Spain) can physically accommodate the relatively few obsolete VLCCs and ULCCs under EU member state flags.

However, most of the docks, although they may presently be underutilised, are used for refits and repairs and their availability for ship recycling is a matter of profitability. The few large docks still engaged in new building in the EU, presumably have a profitable niche production and may not be available for recycling activities.

An extract from Clarksons' demolition database for large EU flagged vessels ready for decommissioning in the years up to 2020 (older than 1995 and >100,000 DWT) shows that 43 vessels presently under a European flag will be recycled (assuming they are not reflagged to non EU member states in that period!). Of the 43 on the list (all tankers) 20 are single hulled tankers to be phased out in 2005 and seven are category 2 tankers, which must be phased out by 2010.

The IMO regulation allows certain exemptions which may allow further trading, but in a worst case scenario 27 vessels will soon await recycling and 12 of these are >200,000 DWT (VLCC) requiring the limited large dry docking capacity. If only the three largest dry docks can accommodate them, it follows that over four years one vessel should be dismantled in each dry dock. Obviously, if both the vessels OWNED by EU registered companies and the vessels FLAGGED in the EU member states must be recycled in the EU or OECD, then the number of vessels competing for a limited dry dock capacity will increase.

Still under the assumption that all vessels will maintain the flag and ownership construction, some 103 vessels are to be scrapped in the EU until 2020 – corresponding to eight vessels per year. However, 55 are single skins that are to be phased out by the end of 2007 unless exempted. Up to 41 will be >200,000 DWT (VLCC), so large that they may only be recycled in the largest dry docks.

Assuming the same capacity in three European yards as planned in the Pipavav facility in India, 12 VLCC can be dismantled per year. This suggests that vessels may have to be laid up, that more facilities should be included or that dry docking should be combined with pre-cleaning and topsides dismantling while moored at a pier.

***Pier breaking accepted***

The claim for dry docks by the NGOs is based on the recommendations of the ESM guideline of the Basel Convention because the text mentions "impermeable surface" as a requirement during dismantling. This requirement is currently interpreted as the ships' hull may serve as an impermeable barrier and in combination with a wet barrier (a boom) the existing methods of pier breaking and slipway breaking do comply. If the pier breaking methods continue to be accepted as green recycling there is a large dormant capacity meeting the physical dimension requirements in ports and waterways throughout the EU, even for very large vessels. In practice, pier breaking are often combined with docking capacity or slipway for the breaking of the keel.

***Pre-cleaning capacity***

Pre-cleaning without scrapping afterwards is not carried out in any EU member state harbour on a regular basis. An example of pre-cleaning is the former UK navy vessel, HMS Scylla, which was decontaminated prior to being sunk off Whitsand Bay near Plymouth to become an artificial reef. In such cases, the London Convention Specific Guidelines for Assessment of Vessels will apply (UK 2006b).

Clearing a vessel for hazardous waste or pre-cleaning it is generally not seen as a market area for the yards since they are typically borne out of the metal recycling business. This task is usually outsourced to specialist companies that may perform most of the activity while the vessel is moored in port rather than on the yard premises. Thus, pre-cleaning of vessels should not obstruct utilising the yard's capacity for scrapping of the hulls, although with larger and more complicated constructions the cleaning may be more intermittent with the actual dismantling. On an *ad hoc* basis the Dutch authorities had the *Sandrien* and the *Otapan* precleaned. Also, the French authorities had the *Clemenceau* pre-cleaned prior to departure for Turkey.

In several EU countries regulations exist as to the requirements for the precleaning of obsolete vessels prior to the use as targets for naval and military shooting practise. Thus, the skills and facilities for pre-cleaning of ships are available, but the scope and intermittent need for the service has not led to the emergence of a viable market. It is clear that pre-cleaning has been an arduous and time consuming process in these cases. Several of the vessels were pre-cleaned for more than 12 months. In yards engaged in the subsequent scrapping it is estimated that pre-cleaning takes 4-7 weeks.

## FACILITIES IN EU CANDIDATE COUNTRIES TURKEY AND CROATIA AND EFTA COUNTRY NORWAY

### The shipbreaking facilities in Aliaga, Turkey

There are some 25 plots for shipbreaking in Aliaga that are operative today. The procedure can be characterised as slipway recycling: there is a considerable draught near the coast and even large vessels of 10,000 LDT and more are run ashore.

The vessel is cut from the bow and sections of 600-800 tonnes are cut and pulled onto the cutting area by wrenches. Smaller sections and equipment are removed by cranes. Vehicles and cranes can operate close to the vessel on the shore or on barges.

The plots are situated on Government property previously leased for only five years at a time. In 2006 the lease terms were changed and leases now runs for 20 years. The longer perspective is considered by the Turkish Shipbreaking Association (TSA) to increase the willingness with yard owners to invest in safety, health and environmental improvements of the yards. The yards were the topic of a critical Greenpeace report (Greenpeace 2002), and since then a hazardous waste programme has been initiated by the TSA, who hired a manager with hazardous waste management responsibilities.

This has led to improved procedures in yards and the programme also increased the availability of facilities for hazardous waste: a mobile asbestos decontamination unit, an organised oily waste collection, and hazardous waste reception facilities are now available to the yards. Some yards (e.g. Cemas) have now large concrete paved areas for the cutting procedures; others have invested in their own asbestos unit and oil pumping and storage facilities not to mention obtaining ISO 9001, ISO 14001 and OHSAS 18001 certification (Leyal Ship Dismantling & Recycling).

The Simsekler yard was chosen by both French and Dutch authorities for the proposed recycling of the *Clemenceau* and the *Otapan*, respectively. Compared to the situation described in the Greenpeace report in 2002, the improvements in several yards in Aliaga are significant.

A compliance with the requirements of the Basel Convention Guideline on Ship Dismantling is not a far fetched possibility (Neser et al 2006), although the acceptance of wet recycling is a prerequisite. The full annual capacity has previously been given as 600,000 LDT with 800- 1200 employees (~500 LDT/ man year), but recent estimates by one of the recycling yards brings the number up to 1,000,000 LDT/year (Ayvatoglu 2006).

Turkey's Shipbreaking Regulation came into force in March 2004 and in early 2007 a re-permitting exercise was carried out for the active yards. Four yards were awarded permits, whereas the remaining yards were granted conditional permits to be renewed after reconsideration.

### Shipyard facilities in Croatia

There are five major shipyards in Croatia organised in a joint company where the Croatian Government holds a majority of shares. There are no reports on ship recycling of larger vessels, but the yards in Brodotrogir and Kraljevica are involved in ship repair activities for vessels up to 50,000 and 25,000 DWT, respectively.

## Upgrading in Turkey and Croatia

### *Turkey*

The options for upgrading facilities in the OECD country Turkey is particularly interesting for the EU, since Turkey is in a pre-accession mode and therefore is underway with adhering to the EU legislation. Thus, if the capacity of the existing shipbreaking facilities in Aliaga close to Izmir meets the appropriate standards, EU vessels can be exported thereto without violating the Basel Convention, the Ban amendment or Waste Shipment Regulation.

The facilities in Aliaga are only accountable as safe recycling facilities if dry docking is not a requirement. The methodology of landing on the shore and pulling the vessel a shore and cut it from the bow may need to be refined. This may be achieved by using pontoon bridges or barges along the vessel to accommodate the use of heavy equipment alongside the vessel and at the same time reducing the wave action on the grounded vessel. It is a possibility to pre-clean vessels in the EU and to tow the emptied hulls to scrapping in Turkey. In this case, however, the transport to a non-OECD country is legal and may be more profitable.

### *Croatia*

The options for using Croatian yards for shipbreaking are similar to the EU and Turkish conditions in terms of the need for adherence to the EU *acquis*. There are no reports on any interest in ship recycling of larger vessels in Croatia.

## EFTA country Norway

Norway has a number of facilities that may enter the ship recycling market for larger vessels. A facility in Norway (Fosen Gjenvinning AS) is active on the demolition market for smaller vessels and may be upgradeable to accommodate larger vessels. No plans have been published. Three facilities in Norway: Aker Stord, Norsk Metallretur Offshore Recycling AS and AF Decom AS are active in the decommissioning market for offshore platforms. None of these have any current plans of entering the ship recycling market.

## GREEN SHIP DISMANTLING FACILITIES OUTSIDE EU AND EUROPEAN OECD COUNTRIES

The companies shown below have identified themselves as ISO 14001 and/or OSHA 18001 certified or they are situated in an OECD country and as such *a priori* acceptable for export of vessels for recycling from the EU. Since Norway and Turkey have been included previously, this would include the OECD countries Australia, Canada, Japan, Korea, Mexico, New Zealand and USA.

## OECD countries

### *USA*

During 2006 there were seven ship recycling facilities operating in the USA located in Brownsville, Texas; New Orleans, Louisiana, Norfolk, Virginia and Baltimore, Maryland. Their clients are MARAD, the U.S. Navy or commercial operators. The approximate total capacity in the U.S. is 225,000 tons per year at the facilities.

The four companies located in Brownsville, Texas can accommodate a total 150,000 tons per year, with International Shipbreaking Limited providing one-third of the total Brownsville capacity. Vessels for recycling have not been imported to the USA in recent years. The USA is not a Party to the Basel Convention, and the import would fall under the Toxic Substances Control Act (TSCA), which also prescribes a pre-notification. The US EPA has stated that an application to import a ship would require a 9 - 12 months public review and approval process and that the license to import the vessel would be limited to one year.26

### *Canada*

In Port Colborne, Ontario, the International Marine Salvage Inc. dismantles Canadian and imported vessels in a facility that is ISO 14001 certified. The company has recycled in excess of 150,000 tons of steel from dismantling ships since 1985. This steel has primarily been re-melted into new steel by domestic steel producer, but substantial tonnage has also been transported to various mills in Asia, for re-rolling into new bars and structural shapes. Canada is a signatory to the Basel Convention.

### *Other OECD countries*

No recycling facilities for larger vessels have been identified in Australia, Japan, South Korea, Mexico, New Zealand or Iceland. Japan and Korea were scrapping nations in the 1970s and 1980s and Mexico had a scrapping industry until the 1990s. The Mexican Industry was partly based on vessels exported from the USA. The countries will, however, typically have marine salvage companies with some access to breaking facilities, but these are typically repair and maintenance yards and only used for scrapping vessels on an *ad hoc* basis. Besides the USA, the only OECD country with any significant ship recycling industry is Turkey.

## Exporting to other OECD countries

Recycling facilities on both the Atlantic and Pacific coasts of the Americas will be useful considering that a merchant vessel may be *en route* or in any given port in these waters when a decision to scrap is taken. At present the majority of active green recycling capacity is located in the USA due to the

US Navy and MARAD demand for ship scrapping under acceptable safety, health and environmental conditions. There is no clear indication that the American facilities may operate on an open international market and become available for foreign vessels for scrap.

The Canadian facility has catered primarily for the domestic market, but has recently expressed interest in operating on the international market in particular regarding government vessels. There is a size limitation to the import of ocean going vessels as they must pass the locks of the Welland Canal.

## ASIA AND NON-OECD COUNTRIES

In the countries typically ranked as prime shipbreaking countries national associations of shipbreakers exist, and some of them or other authorities have initiated activities to upgrade their national ship recycling yards.

This is especially true for China and to some extent India. In both countries some ship recyclers have obtained the quality management ISO 9001 certifications.

### *India*

According to press releases from Gujarat Maritime Board, which is the managing authority of the shipbreaking on Alang Beach, 22 companies in Alang area have received ISO 14001 and/or OSHA 18001 certification. A list of 32 facilities in Alang certified for ISO 18801 and OHSAS 18001 was available from a cash buyer.

Most yards were certified by International Certification Services (Asia) Pvt. Ltd., and some through the classification society RINA. Three yards with a combined annual capacity of 60,000 to 100,000 LDT present their certifications on their company homepages on the internet.

According to the Gujarat Pollution Control Board the yards at Alang now have access to a Treatment, Storage and Disposal Facility (TSDF) for hazardous waste from the ship breaking activities (Gujarat Pollution Control Board 2006).

The Pipavav facility near Alang is a large semi dry dock facility constructed for "green" recycling of up to four VLCCs at a time. It was developed by private Indian funds and financial assistance from Japan in the form of loans. It is presently not active in the recycling market.

### *China*

In the late 1990s P&O Nedlloyd was challenged to improve their decommissioning practices and the company subsequently changed its policy on the matter. They chose two yards near Shanghai and initiated a capacity building project on safe recycling. P&O Nedlloyd ended up recycling 19 vessels in China between 2000 and 2003 under the company's supervision. On the shoulders of this effort, the Chinese National Shipbreaking Association initiated a campaign to upgrade the safety, health and environmental issues at four yards, partly supported by the China – Netherlands Environmental Protection Cooperation Project (2005). Chinese authorities regulated shipbreaking specifically already in 1988, and recently implemented the "General regulations on green ship recycling" and "Technical guideline for Pollution Prevention related to Shipbreaking" from 2005 and 2006, respectively.

At least four yards have achieved ISO 14001 and OSHA 18001, i.e. the certifications regarding management of environment and health and at the recent IMO Regional Seminar and Workshop held 5-8 February 2007 in Zhu Hai, China, the companies capable of recycling large vessels (VLCCs and ULCCs) in China were given as:

- Xin Hui Shuang Shui Ship Breaking & Steel Co. LTD
- Zhongxin Shipbreaking & Steel Co.
- PanYu Ship Breaking and Steel Rolling Co.
- Chang Jiang Ship Breaking Yard.

The capacities have previously been given as <100,000 – 300,000 LDT/year for each yard and collectively amount to >800,000 LDT/year. In the autumn of 2006, China NSA carried out a trial on ship recycling in accordance with the draft convention on ship recycling. The trial took place at Chang Jiang Ship Breaking Yard on the container ship Nedlloyd Dubai (BIMCO 2007). This yard was also used during the first Chinese "green" ship breaking campaign in 2000- 2003.

***Pakistan and Bangladesh***

It has not been possible to identify shipbreakers in Bangladesh or Pakistan with certifications regarding health or environment, such as ISO 14001 and OSHA 18001. The ILO project on ship recycling in Bangladesh did not report on any certified yards (Feringa 2005), but the PHP Group is reportedly seeking certification for a yard in Chittagong. No information is available from Pakistan.

***Other countries***

Japan has supported developments of shipbreaking in the West Cebu in the Philippines and as previously mentioned provided financial assistance to the construction of the Pipavav facility near the Alang area. None of these enterprises have proved viable in the shipbreaking market.

Also, Vietnam is engaged in the shipbreaking market but presently only on a very limited scale (1% of tonnage). No information is available to assess the relevant compliance level of these facilities.

**Projections for Asian yards**

According to press releases from the Gujarat Maritime Board, 22 facilities at Alang have reported compliance to ISO 14001 on environmental management, and information from the cash buyer GMS lists 32 yards with ISO/OHSAS certifications in Alang.

It is emphasised that although the present study is focused on environmental issues, the acceptance of a yard as an operator in safe and environmentally sound ship recycling includes the procedures of safety and human health issues.

The Pipavav Ship Dismantling Engineering Limited may still be considered a dormant recycling facility located at the Pipavav Port in the State of Gujarat some 70 kilometres from Alang Beach. The facility was constructed with the aim of meeting the new standards for environmental protection and was finalised in 2003.

It was built by Sumitomo Corp. of Japan for Gujarat Pipavav Port Ltd. with a loan from Overseas Economic Corp. Fund of Japan to provide a safer recycling than the nearby Alang beach, but has now been converted to a ship building yard. The dimensions and procedures of the Pipavav facility have been described before (COWI/TREN 2004), but in brief it is two sloping wet docks each with a length of 680 meters and widths of 60 and 65 meters, respectively.

The first 70% of the dismantling of a vessel is performed at the deeper end of a dock and the vessel is moved to the shallow dock end during a high tide where the remaining 30% of the dismantling is performed. The docking capacity allows dismantling of four vessels simultaneously.

The annual dismantling capacity under normal conditions is planned to be 275,000 LTD with an optimal capacity of some 400,000 LDT. Unfortunately, no ship has to date been recycled at the facility. According to the IMO MEPC seminar in February 2007 in Zhu Hai, China, the capacity is more than 3,000,000 LDT/year, and it is estimated that 800,000 LDT/year is in yards that have ISO and OSHA certifications or have had procedures upgraded and audited by P&O Nedlloyd.

The latter yard has projected expansions to provide capacity for up to 1,000,000 LDT/year, bringing the total green capacity up to 1500,000 LDT/year.

## SUMMARY OF EXISTING AND PROJECTED CAPACITY AND RECYCLING NEED

The existing capacity of green recycling worldwide may approach 200,000 LDT/year in the EU, but is at present not suitable for vessels of VLCC size and larger.

**Table. Summary of existing and dormant capacity for safe and environmentally sound ship recycling**

| Safe and environmentally sound ship recycling | Existing capacity | Dormant if dry dock required | Dormant capacity by pier or landing method |
|---|---|---|---|
| EU | 200,000 | 1,000,000 | Large |
| Candidate countries (Turkey)** | 50,000 | - | 600,000 – 1,000,000 |
| OECD (US + Can) | 225,000 | Not estimated | Not estimated |
| Asia*** | 800,000 | Not estimated | 1,500,000 – 3,000,000 |
| **Total** | **1,325,000** | **1,000,000** | **1,950,000 – 4,000,000** |

* Assuming three docks with the same combined capacity as the design capacity of the former Pipavav facility in India (400,000 LDT/year) + Able UK with a planned capacity up to 600,000 LDT7year.
** The capacity in Turkey is yet not fully accepted as green recycling.
***Chinese yards. Information from IMO MEPC seminar in Zhu Hai, China suggest expandable to 3,000,000 LDT/year

The above-mentioned Indian yards with certifications would add >1,000,000 LDT/year to the global green capacity. However, with the present knowledge of containment employed in the beaching method, the yards cannot be accepted as green recycling.

**Table. Comparison of existing and dormant safe and environmentally sound recycling capacity with the potential supply of government vessels and merchant vessels**

| **EU Government vessels and fishing vessels** | **LDT/year** |
|---|---|
| Recycling demand | 40,000 |
| Existing EU capacity | 200,000 |
| Balance | +160,000 |
| Capacity need | No extra capacity needed |
| Certified facility in Turkey* | 50,000 |
| **European flagged merchant fleet** | **LDT/year** |
| Recycling demand | 1,600,000 |
| Existing EU | 200,000 |
| Mobilisable in EU (existing infrastructure) | 1,000,000 |
| Balance | -400,000 |
| Capacity need | Extra capacity needed |
| Mobilisable capacity in Turkey* | 600,000 |
| Certified facilities in China* | 1,500,000 |

* The capacity in Turkey and China is not yet fully accepted as safe and environmentally sound recycling and the facilities in China are outside of OECD.

In summary, the existing European capacity will be sufficient for the fishing vessels, navy vessels and other government owned vessels to be recycled over the next 10 years.

To provide green recycling in the EU for the merchant fleet presently under EU member state flag the dormant capacity must be invigorated. The number of potential facilities and the physical dimensions of the potentially available structures are just sufficient to accommodate the maximum sizes of the fleet, but several dry docks must be opened for recycling. It will also be necessary to continue to use pier breaking for pre-cleaning and partial dismantling.

However, it should be kept in mind that the demand for scrapping of 1.6 million LDT/year of EU member states flagged vessels is an average of a range up to 2.1 million LDT/year excluding the year 2010. The demand for recycling of EU member state flagged vessels in the peak phase out year of 2010 is projected to 2.7 million LDT.

If vessels owned by companies in EU member states are included one may add 3.1 million LDT for recycling in 2010. Obviously, capacity cannot be established for a single peak in the demand, but additional capacity will be needed for average years to ensure the access of European merchant ship owners to certifiable green recycling. This additional capacity can to some extent be found in Turkey provided the shore landing methodology is accepted as safe dismantling.

The capacity available or mobilisable in China adds significant volume to the demand that can be met for green recycling. However, since China is not a member of the OECD the yards may at present only come into play for EU-flagged ships that are not exported for scrapping from the Community, i.e. ships that do not begin their final voyage in an EU port.

## OVERVIEW OF STANDARDS AND GUIDELINES FOR SHIP DISMANTLING

This section draws an overview of existing standards of environmentally sound management in ship dismantling. The overview briefly outlines and compares the existing standards with special attention to the methods applied by the standards, also addressing if a certification system and third party accreditation is included in the different standards. The overview further includes Environmental Health and Safety standards (EHS) of international organisations, which could be applied to ship dismantling. The applicability of the guidelines towards controlling EHS-issues in ship dismantling is analysed.

Finally, the overview includes guidance developed by EU member states, other flag and recycling states as well as shipping associations and private shipping lines, for which the existence of company policies directly on ship recycling or environmental policies including ship recycling is described.

## EXISTING STANDARDS AND GUIDELINES ON SAFE AND ENVIRONMENTALLY SOUND MANAGEMENT IN SHIP DISMANTLING

Several international standards and guidelines on safe and environmentally sound recycling of ships have been developed by different international organisations and national governments. The list below includes the identified standards and guidance documents related to the recycling of vessels:

- UN bodies technical guidelines
  - Basel Convention Technical Guidelines for the Environmentally Sound Management of the Full and Partial Dismantling of Ships
  - ILO: Safety and health in shipbreaking: Guidelines for Asian countries and Turkey
  - IMO Guidelines on ship recycling
- Specific guidance on pre-cleaning and decontamination
  - IMO Guidelines for the Development of the Ship Recycling Plan
  - IMO recommendations & guidance on "Gas-free-for-hot-workcertification" during ship recycling operations
- Industry codes and standard contracts
  - ICS Industry Code of Practice on Ship Recycling
  - ICS Inventory of Potentially Hazardous Materials on Board
  - BIMCO Standard Contract for the Sale of Vessels for Demolition and Recycling (Demolishcon 2003)
- Practical manuals on ship dismantling
  - Draft Pocket Manual on Implementation of Green Ship Recycling, prepared by Danish Environment Protection Agency (June 2006)
  - U.S. EPA "A Guide for Ship Scrappers – Tips for Regulatory Compliance".

## IMO, ILO AND BASEL CONVENTION GUIDELINES

The three United Nations bodies: the International Maritime Organisation (IMO), the International Labour Organisation (ILO) and the Conference of Parties to the Basel Convention on the Control of Trans-boundary Movements of Hazardous Wastes (the Basel Convention) are among those who have been engaged in the area for the longest period and produced the most detailed guidelines within the field.

However, none of the three guidelines are mandatory and the practical application of them to date has been limited. In the recognition of the need for coordination of their work with the aim of avoiding duplication of work and overlapping of roles, responsibilities and competencies between the three organizations, a joint working group on ship scrapping was established in 2005.

The work and outcome of the joint working on ship scrapping established by IMO, ILO and the Basel Convention. Besides this joint work IMO have themselves initiated work on developing a binding International Convention for the Safe and Environmentally Sound Recycling of Ships, the IMO Convention on Safe and Environmentally Sound Recycling of Ships.

In parallel to the negotiation of a binding international regime for clean ship dismantling, a number of guidelines are currently being prepared. The guidelines will supplement the IMO Convention. At MEPC 55 the intercessional Correspondence Group was tasked with the development of an outline of one or two pages for each guideline.

Some guidelines are further developed already by the Correspondence Group including the guidelines for ship recycling yards on safe and environmentally sound ship recycling, which is being drafted by Japan31. The primary intentions of this draft guideline are to provide information to operators of ship recycling facilities to enable them to understand how to comply with the ESM requirements of the Convention. The Japanese draft guideline for ship recycling yards on safe and environmentally sound ship recycling details the ESM requirements of the Convention after it has entered into force.

The requirements are specified for the four different methods: beaching, landing, afloat and dry. Specifically regarding beaching, the draft guideline specifies that strict management requirements are necessary and that the method is not recommended for newly-established facilities. An example of a specific requirement for the beaching method is installation of appropriate oil reception and/or storage apparatus adjacent to the beached hull, in order to collect residual oil, solvents used for sludge, and/or oil cleaning. The draft guideline is still under negotiation and its final destiny is not known at the time of writing.

## EXISTING INTERNATIONAL EHS STANDARDS

Besides the above referred international standards and guidance on environmentally sound management in ship recycling as described above, different international standards and guidance on EHS-management exist. The most widespread EHS-standard which can be and is being used to control the environmental impacts of ship recycling is the environmental standard ISO 14001 and the standard for occupational health OHSAS 18001. These standards are described and analysed below. Focus is on their applicability in ship recycling, hereunder features about the standard that needs special attention.

## ISO 14001

The International Organization for Standardization, ISO, is a non governmental organisation that has developed the widespread environmental standard, ISO14000, which consist of a series of voluntary standards and reference documents designed to provide the general framework for organizing the tasks necessary for effective environmental management. Frequently, the series of standards are referred to as ISO 14001, because this is the number of the actual standard.

All other sections in the series are support documents, including guidelines for implementation, environmental auditing, environmental labelling, and product life cycle assessment. The ISO 14001 standard is a generic management system standard, which means that the same standards can be applied to any organization, large or small and whatever its product or the scope of activity is. The ISO 14001 standard must be followed by third-party verification to obtain certification towards the standard.

The first section of the standard directs the organization to identify its activities, products, or services that can or already significantly impact the environment. Secondly, the organization must develop an environmental policy "appropriate to the nature and scale" of the activities, products, or services with significant environmental impacts.

There must be a specific commitment to prevent pollution. Top management must define the policy and ensure that it is communicated to all employees. In the policy, the organization must commit to comply with all pertinent environmental legislation, regulations, and/or legal requirements.

Perhaps most important, the section directs the organization to set, review, and document definable environmental objectives and targets at relevant functions and levels within the organization. In the third section of the standard, the organization must develop a program for achieving its management objectives.

The program will include assigning responsibility for objectives at "each relevant function and level of the organization" and establishing a time frame for their accomplishment. The fourth major section of the standard calls for the implementation and operation of the environmental management program.

This section includes provisions to educate employees or members at each relevant function and level about the organization's environmental management system and their roles and responsibilities in achieving the organization's goals. In addition, the provisions call for training of employees or members to carry out their tasks within the environmental management system.

Finally, within Implementation and Operation, there must be procedures for preventing and mitigating any environmental impacts from accidents and emergency situations. The Checking and Corrective Action section directs the organization to monitor and measure operations with significant environmental impacts on a regular basis. The organization must establish and maintain a documented procedure for evaluating compliance with pertinent environmental laws and regulations. The standard calls for a continuous audit of compliance with and performance of the environmental management system.

Finally, the standard requires that top management review the environmental management system and provide changes to policy, objectives or operations when necessary which should lead to greater efficiency and continual improvement. ISO 14001 does not include a requirement for the firm to make its environmental performance available to the public.

## OHSAS 18001

The OHSAS 18000 series of standards that includes OHSAS 18001 Occupational Health and Safety Management Systems (OHSMS) grew out of the international success of, amongst others, the ISO 14000 series of standards and the need for managing safety in the work environment. This standard was created from the British Standard for Occupational Health and Safety Management Systems BS 8800:1996.

### *Description*

The OHSAS 18001 came into effect in April 1999. Its full title is "Occupational health and safety management systems - specification". There is a second document in the series (OHSAS 18002), which provides guidelines on the implementation of OHSAS 18001. OHSAS 18001 is based partly on the ISO 14001 diagram on plan, do, check and improve. According to the scope of the OHSAS 18001 specification it is applicable to any organization that wishes to:

- establish an OH&S management system to eliminate or minimize risk to employees and other interested parties who may be exposed to OH&S risks associated with its activities;
- implement, maintain and continually improve an OH&S management system;
- assure itself of its conformance with its stated OH&S policy;
- demonstrate such conformance to others;
- seek certification/registration of its OH&S management system by an external organization; or
- make a self-determination and declaration of conformance with this OHSAS specification.

## EMAS

The EU Eco-Management and Audit Scheme (EMAS) is a management tool for companies and other organisations to evaluate, report and improve their environmental performance. The scheme has been available for participation by companies since 1995 (Council Regulation (EEC) No 1836/93 of 29 June 1993) and was originally restricted to companies in industrial sectors. Since 2001 EMAS has been open to all economic sectors including public and private services (Regulation (EC) No 761/2001 of the European Parliament and of the Council of 19 March 2001).

At this point EMAS recognized ISO 14001 as equivalent to the environmental management system required by EMAS. The mentioned revisions to EMAS have made it easier for organisations already certified to ISO 14001 to attain EMAS registration. For these organisations there will be some minor modifications to be made relating to the core elements of ISO 14001 plus some additional steps specific to EMAS. Additional steps for EMAS registration:

- *Initial Environmental Review* - The EMAS regulation requires that an initial environmental review be performed to identify an organisation's environmental aspects. However, when an organisation already has an EMS certified to ISO 14001 it does not need to conduct a formal environmental review when moving on to EMAS implementation, so long as the environmental aspects (including both direct as indirect environmental impacts) as set out in EMAS are fully considered in the certified environmental management system (EMS).
- *Environmental Statement* - The organisation shall prepare an environmental statement, based on the outcome of the EMS performance audit. The environmental statement must openly report the environmental performance of the organisation.
- *Verifying the Environmental Statement and environmental performance* - In order to attain EMAS registration, the Environmental Statement must be independently validated. This process will check that the statement meets the requirements of EMAS and is publicly available.

From the above description it is obvious that the EMAS requires more transparency of the registered organisation than ISO 14001. EMAS requires that the policy, programme, environmental management system and details of the organisations performance are made publicly available as part of the environmental statement.

ISO only requires that the policy is publicly available. Further ISO14001 does not stipulate the extent to which performance must be improved. EMAS specifies that organisations must attempt to "reduce environmental impacts to levels not exceeding these corresponding to economically viable application of best available technology".

Therefore, it is much easier from the outside of a company to control the actual performance of an EMAS organisation compared to one certified according to ISO 14001 Participation under the EMAS scheme is voluntary and extends to public or private organisations operating in the European Union and the European Economic Area (EEA) — Iceland, Liechtenstein, and

Norway. An increasing number of candidate countries are also implementing the scheme in preparation for their accession to the EU. It is presently not possible for organisations in other countries outside the EU to be registered under the EMAS scheme.

However, according to Commission sources, this will probably be changed in the next EMAS revision which is being drafted. The new EMAS is proposed to allow registration outside EU and to reinforce requirements for legal compliance and documentation of performance improvements. The new EMAS is expected to be active from 2010.

**Applicability of standards in ship recycling**

The ISO 14001 standard and to a lesser extent the OHSAS 18001 standard are already being used in the ship scrapping business with recycling yards being certified according to the standards. Both standards are generic and are thus not designed specifically to ship scrapping industries, and some features about the standard are important to keep in mind when evaluating the applicability of the standard.

The EMAS scheme is less relevant because it can only be used for organisations within the European Union and the European Economic Area (EEA) and some EU candidate countries. However, this may be changed in the next revision of the standard. First of all, it is important to bear in mind that ISO 14001 is an environmental management standard and OHSAS 18001 is an occupational health and safety standard. Certification towards ISO 14001 does therefore not include requirements on proper management of the working environment for workers at the site and vice versa for OHSAS 18001. Secondly, it should be stressed that both standards are procedural as opposed to performance standards.

Therefore, by itself certification against the standard does not guarantee a certain high level of performance. Being a procedural standard means that the standards requirements are referring the organisation's management system (procedures etc.) and not the level of performance. The minimum requirement of an ISO 14001 certified organisation is legal compliance to environmental regulations and continuous improvement. The speed and scope of the improvements is however up to the organisation itself.

The above mentioned requirement of the standards about the regulatory compliance is another important issue, when evaluating the applicability of them towards ensuring sound environmental management within ship scrapping businesses. Many of the large ship scrapping businesses are located in parts of the world where the EHS legislation and regulatory framework are not fully developed, which means that EHS regulatory compliance of a certified company in for instance Asia can easily allow for unacceptable EHS performance according to for instance European environmental legislative requirements.

Finally, a very important issue is the fact that both ISO 14001 and OHSAS 18001 allow for an organisation to certify only part of its activities. Because of this, a firm can choose to bring only a part of its enterprise and business activities under the regime of the standard, thereby leaving processes that are more difficult to manage and improve from an environmental perspective outside the scope of its certificate.

Such proceedings can lead to serious misunderstandings about the value of the environmental management system. Therefore it is not enough to check if an organisation is certified according to these standards. It is also necessary to check the actual certificate to see which part of the organisation is covered, and possibly to check the documents behind to fully understand the scope of the certification.

This information is, however, often not available as the standards (except EMAS) only requires the policy to be public available. EMAS requires that the policy, programme, environmental management system and details of the organisation's performance are made public available as part of the environmental statement.

Further the statement is verified by an external third part certified auditor. Therefore it is always possible to control the environmental performance of an EMAS registered company. A certified management system does not guarantee a certain level of performance except from legal compliance. Therefore it is necessary to establish documentation of best practise/ best available technology. With this in hand it will be possible to evaluate the performance of a ship-recycling company.

Such documentation can be used by both local authorities in the permitting process and by ship owners seeking a proper recycling facility. A certified management system is valuable as it sets the framework for an organisation's environmental or health and safety work, but seen from the outside it needs to include a requirement for transparent reporting (like the EMAS scheme).

And when it comes to evaluation of performance it is also needed to have documentation of best practise/ best available technology. Further, it will be valuable to have guidelines for local authorities on how to handle permitting and inspection and guidelines for ship owners on how to handle the recycling process.

**Existing certifications of recycling yards**

As mentioned a number of ship recycling facilities are apparently already certified according to one of the above-mentioned standards. There is however no combined international register of certified companies according to ISO 14001 or OHSAS 18001.

A register of EMAS registered organisations exists on the EMAS helpdesk but no ship recycling yards were found. It should be noted that the list is nowhere comprehensive. For several of the above mentioned yards more details on the coverage, contents etc. of the certificates were sought together with information of the actual environmental performance of the ISO 14001 certified companies.

Such information was however not available, which just underlines the difficulties of really evaluating the level of performance of the companies and thus the implications of the certificates.

## EUROPEAN UNION GUIDANCE

The EU Waste Shipment Regulation specifically mentions vessels in its Annex III as acceptable to export once 'properly emptied':

- Vessels and other floating structures for breaking up, properly emptied of any cargo and other materials arising from the operation of the vessel which may have been classified as a dangerous substance or waste.

There is no EU guidance related to the cleaning of vessels for export, but several member states have initiated activities on this issue, notably the Netherlands, France, Denmark and the United Kingdom.

### The Netherlands

The Netherlands have obtained practical experience with recycling of contaminated ships and pre-cleaning of ships as a result of the two ships *Sandrien* and *Otapan* being left/arrested in the Port of Amsterdam and taken over by the Dutch Government.

### France

In preparation for the planned export of the former aircraft carrier *Clemanceau* pre-cleaning was carried out by French contractors in the Port of Toulon:

- "All the necessary technical and industrial means were implemented to guarantee the respect of the objectives set by International and European environmental regulations, even though there are no European or International binding regulations regarding ship clean-up or dismantling. The dismantling yard is also certified with ISO 14001 on environmental management systems."

The efforts concentrated on asbestos, sludge, oil and radioactive materials:

- "Prior to the dismantling process, the former Clemenceau's hull was processed and does not contain anymore:
  - Hydrocarbons: all parts of the aircraft carrier with existing or previous hydrocarbons were leached and had their tanks emptied between October 1997 and January 1998. The concerned parts were then filled with freshwater.
  - Polychlorobiphenyle (PCB)/pyralene. Like all the transformers

on French Navy ships, the former Clemenceau transformers were "dry" transformers meaning that they did not contain such substances.

– The 1,238 indicative radio luminescent plates on the former Clemenceau were all removed before the beginning of the asbestos removal works."

Procedures and standards for pre-cleaning are embedded in environmental legislation and no specific procedure addressing this has been developed. As mentioned earlier, a significant effort has been made by the Interministerial Working Group on identifying the need and capacity of European ship scrapping. The planned dismantling of a naval vessel *Lucifer II* in 2007 will take place in Cherbourg at a site where obsolete nuclear submarines are already dismantled (Assemblée Nationale 2007).

**Denmark**

The Danish EPA have been active in ship recycling for a number of years. Their work has included investigation of available ship scrapping capacity within OECD-countries.

The latest initiative includes development of a draft brief "pocket-book" version of the existing *Technical Guidelines for the Environmentally Sound Management of the Full and Partial Dismantling of Ships* published by the Basel Convention in 2003. This pocket book primarily aimed at the operators of yards interested in entering the market for green recycling through upgrading of an existing yard.

The information in the book is given at the level where the management and their contractors can develop the actual plan for upgrading based on the guidance and the information sources given. The pocket book does not in itself include a certification system and third party accreditation, but is instead a step towards development of a standard for EHS certification of recycling yards.

**United Kingdom**

The UK Government was faced with the issue of ship recycling in 2003, when the US Department of Transportation Maritime Administration (MARAD) arranged to recycle 13 of its redundant naval vessels at a Teesside facility (Able Ltd.), and the public and local authorities challenged the status of the permissions at the facility. In the wake of the Ghost fleet issue, the UK published a Strategy and a Consultation Paper and carried out an inventory of ship dismantling capacity in the UK. With respect to Government owned vessels, the Strategy and Consultation paper emphasises (UK Defra 2007) that when "Government owned vessels are either sold for further use to other Governments or commercial entities, or recycled, the following commitments will apply:

- the sales contract for a vessel destined for further use will contain conditions to ensure that the new owner takes full responsibility for the safe and environmentally sound recycling of the vessel;
- recycling will only be allowed in a facility in an OECD country and in accordance with acceptable environmental, health and safety standards; and,
- vessels exported outside the UK for recycling will be subject to the Basel Convention principles of prior-informed consent and environmentally sound management."

The second document is an Overview of the Ship Recycling Process in the UK. It sets out the technical and regulatory requirements for operating sustainable ship recycling facilities in the UK. In addition, a non-exhaustive list of potential sources of assistance, including funding, for those wishing to engage in the industry is provided.

The UK Government's view is that existing legislation on waste does cover ship recycling regardless of whether the ship sails under own power or has to be towed. The Strategy requires any ship sent for recycling from the UK to be recycled at a yard that operates according to standards of environmentally sound management.

To this end the Strategy is available in five languages in addition to English There are three key provisions relevant to ship owners outlined in the Strategy.

Firstly, any UK-flagged vessel being exported for recycling from a UK port will have to be notified according to the requirements set out in the WSR as described above.

Secondly, all import of vessels into the UK for recycling will also be subject to the provisions of the Waste Shipment Regulations (or existing international rules).

Finally, UK ship owners are encouraged to refer to the IMO Guidelines on Ship Recycling at all stages when a ship is sold for recycling.

## GUIDANCE IN OTHER NATIONS

Most of the large ship scrapping nations have developed or are working on developing specific guidelines or standards describing procedures for scrapping of ships within their country. Below are presented the identified national guidelines.

### India

The Central Pollution Control Board in Delhi has prepared environmental guidelines for ship scrapping industries aiming to minimize the effect of the industries on the surrounding environment through proper location of industries and by preparing and implementing an environmental management plan and a disaster management plan.

The guidelines include a description of the appropriate pollution control measures regarding solid waste, air pollution, water pollution and noise. This also includes aspects of workers' safety.

Since 2000, the Gujarat Maritime Board, GMB, has issued regulations covering safety measures for the beaching of vessels. Substance of the regulation was the documentation of a gas-free certificate before beaching and a permission to start the cutting operations to be issued by GMB following the removal of hazardous materials from the vessel. The gas-free certificate was introduced in 2001.

Apart from the third-party gas-free certificates no certification system or third party accreditation exists. The Gujarat Maritime Board has issued Conditions and Procedures for Granting Permissions for Utilising Ship Recycling Plots, dated 30 September 2006. Schedule V (Miscellaneous) contains the following condition: "(6) Permission holder shall use modern technology for recycling activity like use of long nozzle cutting torch, mechanised hydraulic lifting equipments, use of gas detectors, mechanized blowers, classified cutting equipments etc. and which may be prescribed by Gujarat Maritime Board from time to time and according to the norms prescribed by Basal Convention shall have to be strictly observed" (GMB 2006).

**China**

The State Environmental Protection Administration (SEPA) has published a technical manual on preventing pollution from ship demolition. SEPA has initiated work on registration of the Chinese ship scrapping facilities with the objectives to rank these according to their environmental status. The ranking will, reportedly, be linked to ship scrapping licences.

Furthermore, SEPA is about to set up a school on ship scrapping (Blankestijn, 2004). In 1991, the China National Ship Recycling Association (CNSA) was established to take up the responsibility of organizing, managing, co-ordinating and serving this trade, and assist SEPA to observe and supervise the pollution problems arising from ship recycling. CNSA have drafted "General regulations on green ship recycling" (available only in Chinese), which has been in force since June 2005.

In the autumn of 2006, CNSA carried out a trial on ship recycling in accordance with the draft IMO convention on ship recycling. The trial took place at Chang Jiang Ship Breaking Yard on the container ship *Nedlloyd Dubai*. The objectives of the trial was to find out the differences between Chinese ship recycling industry standards and international standards, and to evaluate the affects the convention brought to the Chinese ship recycling industry.

Furthermore, the intention was to study preparatory measures, and to improve occupational safety and health as well as environment protection in the ship recycling industry (BIMCO 2007). CNSA prepared a detailed scheme for the trial, with the draft convention as amended by MEPC 54 as basis, and the relevant ship recycling guidelines as reference. According to recent information (BIMCO 2007) the trial was considered a success and it was concluded that the recycling operation of the vessel essentially complied with the relevant guidelines and the draft convention - mainly because of the yard being capable to do this, and the ship owner keeping strict superintendence of the whole recycling process.

Based on observations from delegates on site for the trial, there is no reason to believe that Chinese yards will have difficulties complying with the (expected) requirements of the IMO convention – also in the interim period from the approval until entering into force (BIMCO 2007). China National Ship breaking Corporation Jiangyin Changjiang Xiagang Ship breaking Company is the yard upgraded and used by P&O Nedlloyd. It has documented procedures and formal certification.

**Bangladesh**

A national regulatory framework for managing ship scrapping is currently to be established. This reportedly includes an approval procedure for the site operator under the responsibility of the Ministry of Commerce and Industry, issuing a "berthing certificate" for each individual vessel to be scrapped and a "hot work" certificate. Presently, however, not all yards are registered with this ministry.

Further, there is a provision under the Environmental Law of 1997 requiring that each and every industry including that of ship breaking must have an environmental clearance certificate from the Department of the Environment, Ministry of Forest and Environment. To achieve this, the ship breaking site must establish an environmental management plan. The Environmental Law is also to cover safety measures, occupational health, waste management, monitoring programmes and disaster management.

ILO has performed a project on Safe and Environment-friendly Ship Recycling (SAFEREC) in Bangladesh (ILO 2003). According to this project there is very little implementation of the above-mentioned regulation in the yards and no enforcement of the "hot work" requirement.

**United States of America**

The United States Environmental Protection Agency, US EPA has developed a set of guidelines for ship scrapping in the United States. The guidelines, 'A Guide for Ship Scrappers: Tips for Regulatory Compliance' were issued in the summer of 2000 (US EPA, 2000). The guidelines are a compliance assistance tool, intended to provide site supervisors at ship scrapping facilities with an overview of the most pertinent environmental and workers health and safety requirements to assist them in ensuring compliance at their facility.

OSHA, the US Occupational Safety and Health Administration obviously also has regulations applicable to ship recycling and works together with the EPA on the issue.

Six active ship recycling facilities exist in the US in 2006. These are approved by MARAD to recycle their old vessels of which more than 100 are from the so-called ghost fleet. These facilities are apparently not open to foreign vessels, but the yards do pursue this issue with the authorities.

## RECYCLING POLICIES AND STRATEGIES OF PRIVATE SHIPPING ASSOCIATIONS AND COMPANIES

### International Chamber of Shipping, ICS

The International Chamber of Shipping, ICS is the international trade association for merchant ship operators. In February 1999, the Industry Working Party on Ship Recycling, IWPSR was established under the co-ordination of ICS. IWPSR published their preliminary Industry Code of Practice on Ship Recycling (ICPSR) in August 2001 (Marisec, 2001). The focus of ICPSR is on issues related to preparation of the ship for recycling, with a primary focus on hazardous materials onboard the ship and the registration and minimisation of the presence of these before delivery for recycling.

Of the more tangible contents of ICPSR is an inventory of potentially hazardous materials on board a ship, consisting of two lists, which have later been included in the IMO guidelines. The ICPSR has no statutory effects towards ship owners, but the industry organisations involved in the development of the code are committed to encourage the widespread use of it within the industry.

The code of practise does not include a certification system or third party accreditation. ICS leads the broadly based inter-industry working group on ship recycling. Their opinion is that some areas of the existing IMO Code on ship recycling should be made mandatory - for example, gas freeing of compartments prior to delivery at recycling yards and the maintenance of inventories of hazardous materials (green passports).

Moreover the industry supports establishment of a system of approved recycling yards, demonstrating which facilities meet acceptable safety standards, to which reference can be made when negotiating the disposal of ships (ICS 2007).

### Baltic and International Maritime Council, BIMCO

The Baltic and International Maritime Council (BIMCO) is the world's largest private shipping organisation with its owner members controlling 65% of the world merchant fleet. BIMCO has recently released their policy on ship recycling (BIMCO 2007). This policy expresses that BIMCO believes that ship recycling is the most environmentally friendly way of disposing of ships and if properly handled BIMCO sees the ship recycling industry as a "green industry". BIMCO's policy expresses the support to the implementation of a legally binding IMO instrument on ship recycling as the most appropriate solution for ship recycling issues, taking into consideration views of the Basel Convention and ILO.

BIMCO's policy further encourages the industry to use the DEMOLISHCON contract when selling ships for recycling. BIMCO's policy expresses the view that it is mainly for the authorities of the flag states to certify ships ready for recycling.

**Intertanko**

Intertanko is the International Association of Independent Tanker Owners. The organisation has 250 members, whose combined fleet comprises more than 2,500 tankers totalling 210 million DWT, which is 70% of the world's independent tanker fleet. Intertanko has in September 2006 announced the adoption of an "Interim Strategy", according to which ship owners should only use recycling facilities that have made demonstrable advances in terms of safety and environmental management requirements as established in Section C of the draft IMO Convention.

**Private companies**

An increasing number of private ship-owners are getting involved in the work on ensuring environmental sound and safe recycling of their ships. A number of companies have prepared inventories of hazardous materials in (some of) their existing ships and new-builds, green passports. The list below includes ships for which green passports have been prepared. The list has been prepared from searches on the internet and other ready available data sources and is not expected to be comprehensive:

- 25 LNG Carriers managed by Shell International Trading and Shipping Company Ltd.
- Aida (Car Carrier, Wallenius Lines)
- Arctic Discoverer (LNG Carrier, K-Line)
- One existing vessel in every class every year (BP Shipping)
- 3 Container ships (Maersk Line Ltd.)
- 3 P-Max Tankers (Concordia Maritime)
- Stena Seatrader (Ro-Pax, Stena Lines).

Regarding ship recycling BP Shipping has developed standards of its own for ship recycling that include the removal of as much hazardous material as possible from vessels before recycling and auditing recycling facilities using BP group health, safety and environmental criteria.

BP Shipping auditors normally witness the recycling process and view the conditions of the facility prior to their use as well as visiting secondary locations such as asbestos disposal sites. All recycling facilities are audited at least every three years. When necessary, as in the recycling of four BP Shipping tankers in Spain in 2004, a BP Shipping Shipyard Health Safety and Environment Superintendent oversees the actual deconstruction of a vessel (BP Shipping 2006).

The shipping company P&O Nedlloyd (now part of Maersk Line) have taken another approach in forming a partnership with two Chinese ship scrapping facilities whose environmental and safety standards were upgraded via technical assistance and training. The partnership includes an agreement on scrapping of the companies' ships against common agreed HSE-standards and procedures.

P&O Nedlloyd initially scrapped 19 ships at these two Chinese facilities in 2000-2003 and now as a part of the Maersk Line continues to use one of the facilities for their recycling demand. Proposals for new procedures and policies leading to safer dismantling are in particular part of the work of the EU financed project SHIPDISMANTL, but also the SAFETOW and SHIPMATES studies address related technical issues. Improvements and procedures have also been published in the technical literature for Turkey (Neser et al 2006) and South Asia (Wijngaarden 2006).

## END-OF-LIFE SHIPS: HAZARDOUS WASTE OR PRECLEANED?

In international waste shipment law it is recognized that a ship may become waste as defined in Article 2 of the Basel Convention and that at the same time it may be defined as a ship under other international rules. The definition of waste in European Community law also applies to ships sent for dismantling.

According to the law waste means any substance or object in the categories set out in Annex I to the Waste Framework Directive which the holder discards or intends or is required to discard.

If a vessel contains considerable quantities of hazardous substances, or in the words of the relevant waste entry GC 030, has not been "properly emptied" of hazardous materials, it will be considered as hazardous waste. The export of such a vessel for scrapping from the EC to a non-OECD country is prohibited under the Basel Ban amendment and the Waste Shipment Regulation, and any dismantling must take place under safe and environmentally sound conditions in an OECD country.

The alternative is to have the ship decontaminated (pre-cleaned) in such a way that the ship no longer constitutes a hazardous waste. This legal interpretation was confirmed by the French *Conseil d'Etat* in the case of the former aircraft carrier *Clemenceau*.

Ships are thus pursuant to Annex II of the Waste Shipment Regulation considered non-hazardous if they fall under the entry GC 030 'vessels and other floating structures for breaking up, properly emptied of any cargo and other materials arising from the operation of the vessel, which may have been classified as dangerous substance or waste'. What the term "properly emptied" means is not precisely defined in the Community Regulation.

We are in the following assessing both legally and technically how the regulation term "properly emptied" may be interpreted, both in relation to cargo and other materials arising from operation of the ship as well as hazardous materials in the ship's structure and equipment.

## COMMUNITY LAW ON HAZARDOUS WASTE

Under Community legislation hazardous waste is defined as "wastes featuring on a list to be drawn up in accordance with procedures laid down in Directive 75/442/EEC on the basis of Annexes I and II of Directive 91/689/EEC." These wastes must have one or more of the properties listed in Annex III of the same Directive.

The most recent version of that list of hazardous waste is included in Commission Decision 2000/532/EC as last amended by Council Decision 2001/573/EC. Ships often contain hazardous materials, which can endanger the environment, if they are not managed in an environmentally sound way. These hazardous materials are listed as hazardous entries in the EU waste list as included in Commission Decision 2000/532/EC as amended.

Annex V of the Waste Shipment Regulation is relevant when determining if a shipment of waste is subject to the ban in Article 16 of the Waste Shipment Regulation.

Article 16 of the Regulation bans waste listed in Annex V if destined for recovery in countries to which the OECD decision does not apply. In Annex V the following hazardous entries relevant for ships destined. Table contains substances, which are likely to be found in relatively large amounts in ships for scrapping, and which is hazardous according to the above mentioned regulation. The substances in the table is based on the list of hazardous materials most often found in vessels prepared by the ICS as part of their 2001 ship recycling guidelines/ICS 2001/.

The table includes information of the hazardous properties of the compounds and if such exist, concentration limits for the hazardous compound. Further information on the Basel convention classification is included. It can be observed that the concentration limits for hazardous materials ranges from 0.1% (1 g/kg) for e.g. asbestos to 3% (30 g/kg) for e.g. tin in tin based anti-fouling coating. PCB has an exceptionally low limit of 50 mg/kg equal to 0.05‰.

For a number of substances concentration limits do not exist, e.g. substances dangerous to the environment. The EU hazardous waste classification requires that the sum of the relative concentrations of hazardous substances compared to the concentration limits should be less than 100%. Unfortunately, as mentioned above, concentration limits do not exist for several of the hazardous waste fractions which are present in relatively large quantities in ships, e.g. oil emulsions and other oil containing wastes.

**Table. Different types of Hazardous waste to be identified at ship, classification of properties and their concentration limits in arelevent entity**

| Substance | Classification of hazardous properties according to EU Hazardous Waste Classification | Description of concentration limits for hazardous waste | Basel Convention Hazard Class, Annex III | Basel Convention, Annex VIII |
|---|---|---|---|---|
| Halons | Depends on the type of halons: Carbontetrachloride: T, Carc3 Methylchlroide: Fx | Yes if conc. > 3 % | H2 | |
| Refrigerants such as R22/R12 | | Regulated according to Regulation 2037/2000 | H2 | |
| Fuel oil, diesel oil and gas oil | Carc. 3 | Yes of conc.>1 % | H3 | |
| Radioactive materials | Radioactive | Dependent on the type of radioactive material | H7 | |
| Waste lead acid batteries | C (R35) | Yes if con. > 1 % | H8 | A1160 |
| Asbestos | Carc1 | Yes if con. > 0.1 % | H11 | A2050 |
| PCB and PCT containing substances | Dangerous to the environment | Yes, if con. > 0.005 % | H11 | A3180 |
| Tin based anti-fouling coatings | Toxic and dangerous to the environment | T: if conc.> 3% | H12 | A4030 |
| Lubricating oils | Dependent on the type of oil but mainly Carc 2 | Yes if con. >0.1 % | H12 | A3020, A4060 |
| Hydraulic Oils | Dependent on the type of oil but may be dangerous to the environment | No concentration limits | H12 | A3020, A4060 |
| Oil residues (sludge), oil water mixtures, waste oils, oil cont. waste | Dependent on the type of oil but may be dangerous to the environment | No concentration limits | H 12 | |
| Polyvinyl Chloride | None | Not hazardous waste | H13 | |

## PRE-CLEANING

The pre-cleaning concept essentially means removal of hazardous materials in the EU and consecutive shipping of the cleaned vessel to a ship scrapping location elsewhere, e.g. in Asia or Turkey. It has been proposed and discussed as a means of solving the EHS issues in the ship scrapping business.

This concept would result in the hazardous materials being removed in Europe, where technologies and facilities for safe removal and treatment of the hazardous materials exist and at the same time allow the major part of the work intensive ship breaking to be performed in Asia, where the labour costs are much lower than in Europe.

The feasibility of the pre-cleaning concept has however been questioned from various sides, both from a technical, safety and economic angle. Some of the arguments against the concept are related to the fact that the pre-cleaning concept includes towing, as gradually removal of hazardous materials from a ship at one point will make the remaining structure not ship-shape (unable to sail for own power). Removal of materials from the ship beyond this point will result in the ship's having to be towed when transported from Europe to Asia. The core of the feasibility discussion can be synthesized into the following scenarios and their internal location on the "pre-cleaning ladder" as illustrated in Figure below:

- Pre-cleaning to a content where the ship is still ship-shape (self-propelled)
- Pre-cleaning to a content satisfying the "proper and necessary cleaning measures" of the EU waste law
- Pre-cleaning to totally remove all hazardous materials in the ship.

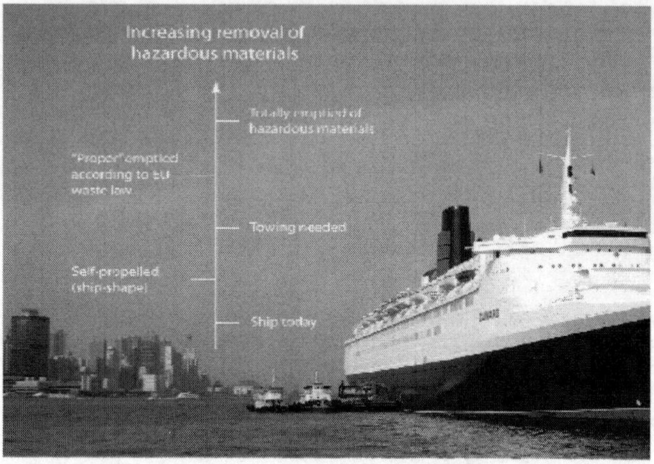


**Fig.** "Ship -cleaning ladder".

**"Ship-shape" condition**

Ships have to fulfil the current safety requirements to be issued with proper certificates to be allowed to sail under their own power. The SOLAS Convention (IMO 1974) is generally regarded as the most important of all international treaties concerning the safety of merchant ships. The SOLAS Convention specifies minimum standards for the construction, equipment and operation of ships, compatible with their safety. Flag States are responsible for ensuring that ships under their flag comply with its requirements, and a number of certificates are prescribed in the Convention as proof that this has been done. Control provisions also allow Contracting Governments to inspect ships of other Contracting States if there are clear grounds for believing that the ship and its equipment do not substantially comply with the requirements of the Convention - this procedure is known as port State control. Examples of the safety requirements in the Convention:

- subdivision of passenger ships into watertight compartments must be such that after assumed damage to the ship's hull the vessel will remain afloat and stable;
- watertight integrity and bilge pumping arrangements for passenger ships;
- stability requirements for both passenger and cargo ships;
- fire safety provisions for all ships and specific measures for passenger ships, cargo ships and tankers, including among others:

— requirements for division of the ship into main and vertical zones by thermal and structural boundaries;
— separation of accommodation spaces from the remainder of the ship by thermal and structural boundaries;
— restricted use of combustible materials;
— ready availability of fire-extinguishing appliances.

How many hazardous materials can be removed from a ship while still keeping it in a ship-shape condition depends on the actual case. It will in general be possible to replace one hazardous material/component with another non-hazardous material/component with the same function. This could in theory mean replacing fire-retardant asbestos with another non-hazardous fire-retardant material, e.g. glass wool.

It will however not be possible to replace fire-retardant material with fire-fighting equipment (EMSA 2007). Special cases exist where it is possible to change the status of a ship, e.g. a passenger ship changing status to for instance cargo ship before the last journey. For this it will be possible to pre-clean for instance the passenger sections of the ship, e.g. remove the fire-retarding asbestos from the passenger cabins and close off these sections (EMSA 2007).

An often discussed waste fraction in relation to pre-cleaning is paint. As long as paint has no safety function like for instance fire-retarding paint, it can be removed and still keep the ship in ship-shape condition. This includes paint on the outside hull, but also antifouling paint under the waterline. Pre-cleaning of a ship to a level where it is still ship-shape will in most cases not be sufficient to meet the EU waste shipment law's requirement of "properly emptied ". This can most easily be seen from the fact that oil, including bunker oil, has to be removed from the ship, which renders the ship unable to sail under its own power.

**Towing**

Whenever material or components have been removed from a ship, which have made it unable to meet the safety requirements discussed above, the pre-cleaned ship will have to be towed to the final breaking location. No international guidelines exist for approving ships for tow. Such ships need to be approved by a recognised classification society who checks the safety conditions of the ship in relation to the proposed tow (EMSA 2007). E.g. in Denmark the following requirements need to be fulfilled by the precleaned ship (similar requirements exist internationally):

- Fulfilment of load line (stability calculations/test);
- Mounting of towing arrangements, including spare arrangements;
- Mounting of appropriate lanterns.

From a technical point of view almost all material/components can be removed from a ship while still keeping it in a towable condition as long as parts of the primary ship structure are not removed. This means for instance that the main deck cannot be removed (EMSA 2007). Towing of pre-cleaned ships will be associated with a different risk-pattern than sailing of ships for its own power.

In cases where no hazardous materials are located inside the primary structure of the ship, a total removal of all hazardous materials is possible. Between the two extremes, the ship as it is today and the totally pre-cleaned ship, are several intermediate situations, including the situation of removing as much as possible of the hazardous materials in order to fulfil the "properly emptied" requirement of EU waste law.

**"Properly emptied"**

It can be observed that EU waste law includes concentrations ranging from 0.1% (1 g/kg) for e.g. asbestos to 3% (30 g/kg) for e.g. tin based anti-fouling coating. PCB has an exceptionally low limit of 50 mg/kg equal to 0.05‰. For a number of substances concentration limits do not exist, including substances dangerous to the environment.

For compounds where concentration limits exist EU waste law does not specify what material fraction the hazardous compounds shall be seen in relation to. This is important information in order to be able to calculate the concentration of the compound in the material fraction and from that evaluate if it is to be considered as hazardous waste. If this ship itself as a whole has to be classified as hazardous waste based on the total amount of asbestos or the total amount of tin-based antifouling coating, the total amount has to be above 15.4 tons or 463 tons, respectively.

These amounts of the two compounds are required to bring the total concentrations above the concentration limits to classify the waste as hazardous. The above example clearly shows that the volume of hazardous materials should not be evaluated against the total ship weight. Instead it is appropriate to evaluate the amount of hazardous material against the smallest material fraction, which can be taken out of the ship as a separate fraction. Such smallest separate material fractions have to be decided on a case-by-case basis. Examples could include:

- The total exterior paint layer on the outside of the ship hull (could consist of several different paint layers, but such cannot be separated);
- Total wall panels containing asbestos;
- The PCB-containing oil in transformers (not the total weight of the transformer).

For compounds where no concentration limit exists, the smallest separate material fraction containing the hazardous compound must be removed as hazardous waste. From a practical point of view a maximum allowable threshold concentration limit should be defined for compounds where no concentration limits exist. The limit could for instance be relevant in a case where for instance oil tanks are emptied of oil, but these still contain traces of oil. Such limit could be defined at 0.1%, which is the concentration limit for the most dangerous compounds (except for PCB, where the concentration limit is 0.05 ‰).

**Technical feasibility**

As stated in Section 6.3.2, the theoretical upper limit for pre-cleaning of vessels consists in removing parts of the primary vessel structure, as this will leave the remaining structure unfit for towing. In practice a lower limit for pre-cleaning will exist, where pre-cleaning above this limit most likely will be unfeasible.

The "location" of this practical limit for pre-cleaning will have to be established on a case-by-case basis, but some general considerations on the location can be established. The feasibility of precleaning will be reduced whenever planning for removal of difficult accessible hazardous materials, which will include breaking-up of other non-hazardous structures to get access to the hazardous materials for removal.

The more of these operations have to be done to pre-clean the vessel under consideration, the closer one comes to the limit where it will be easier for the ship owner to break-up the entire structure at the "pre-cleaning facility". A more pronounced case of these unfeasible pre-cleaning measures are measures that involve breaking-up of structures on the vessel that will have to be replaced afterwards to enable the remaining vessel to be in a towable condition. Such measures will be costly and reduce the feasibility of pre-cleaning of the vessel quite fast.

Pre-cleaning measures that result in replacement of other structures include: removal of large structures from the inner sections of the vessel, which will result in the need for cutting holes in the outer structures that will have to be closed again afterwards, and removal of structures that stabilise other structures which also have to be removed. The main engine is an example of a large structure that most likely will be unfeasible to remove. A lower deck supporting or stabilising upper decks is another example of a structure which will be unfeasible to remove from a stability/structural point of view. The location of this practical feasibility level for pre-cleaning on the "precleaning ladder" has to be decided on a case-to-case basis.

Asbestos has often been used for fire sectioning structures including decks, which is why a location below the "properly emptied" level is expected to be rather common. Another technical issue that is affecting the feasibility of the pre-cleaning concept is the availability of appropriate towing vessels. Based on the projection of future scrapping requirement an estimate of up to 200 pre-cleaned vessels a year, to be towed from Europe, can be considered.

If all these vessels have to be towed to Asia, it will require around 66 tug-vessels working fulltime. Such tug demand will result in temporary shortage of tugs, until the demand has been met by the supply of new tugs.

## EXISTING PRE-CLEANING GUIDELINES

Other organisations or governments have developed guidelines for pre-cleaning of ships. These guidelines are not developed to cover scrapping of ships, but can provide input to the discussion of appropriate cleaning measures. Some of these guidelines are described below.

### US artificial reefing of vessels

The US Code of Federal Regulations, Title 40, Volume 12, Parts 190 to 259 as revised July 1, 1996, permits in section 229.1 the burial at sea of vessels and airplanes registered in the United States or flying the United States flag under certain conditions, including: "appropriate measures shall be taken, prior to disposal, by qualified personnel to remove to the maximum extent practicable all materials which may degrade the marine environment, including without limitation:

- emptying of all fuel lines and fuel tanks to the lowest point practicable, flushing of such lines and tanks with water, and again emptying such lines and tanks to the lowest point practicable so that such lines and tanks are essentially free of petroleum, and
- removing from the hulls other pollutants and all readily detachable material capable of creating debris or contributing to chemical pollution".

### OSPAR Artificial Reefs Guidelines

The Oslo Paris Convention for protection of the North Atlantic (OSPAR) has also developed guidelines on artificial reefs within their territory. These guidelines specify that artificial reefs should only be established if, after due consideration of all socio-economic and environmental costs (e.g. undesirable impacts or alteration), a net benefit can be demonstrated, in relation to the defined objectives. In such assessment of potential effects (which may have to be a formal environmental impact assessment if major impacts cannot be ruled out) the following steps should be followed:

- Studies should be carried out that yield the information required to assess:
  - possible impacts of the installation of an artificial reef on the indigenous fauna and flora and the environment of the site and the wider surroundings;

In relation to the requirements for the materials for use as artificial reefs the OSPAR Guidelines specify:

- Artificial reefs should be built from inert materials. For the purpose of these guidelines, inert materials are those which do not cause pollution through leaching, physical or chemical weathering and/or biological activity. Physical or chemical weathering of structures may result in increased exposures for sensitive organisms to contaminants and lead to adverse environmental effects.
- Materials used for the construction of permanent artificial reefs will of necessity be bulky in nature, for example geological material (i.e. rock), concrete or steel.
- No materials should be used for the construction of artificial reefs which constitute wastes or other matter whose disposal at sea is otherwise prohibited.

The Governments of France and the United Kingdom have developed guidelines to comply with the OSPAR recommendations for cleaning of vessels to be used as shooting targets or for artificial reefs.

## RISK ASSESSMENT OF TOWING OF PRE-CLEANED SHIPS FROM EUROPE TO ASIA

Pre-cleaning of ships is closely connected to towing as the pre-cleaned remaining structure often will be in a shape unfit for sailing under its own power and which then has to be towed from the precleaning location to the final scrapping location. The objective of this section is to assess the risks of towing of pre-cleaned ships from Europe to Asia. To meet the objective, a quantitative assessment has been made of the risk of total loss of a towed ship in connection with towing it from London in England to Alang in India.

Two routes are considered, one through the Suez Canal and another around the Cape of Good Hope. The routes are defined from the "portworld" web site. The two locations have been chosen as representatives of a European port and a major Asian scrapping location, respectively. Changes to departure harbours within Europe and destinations within Asia will affect the results only marginally.

### Risk assessment

#### *Towing of ships*

Towing is a routine operation for tugs. Good practice of seamanship is necessary to accomplish the mission without endangering the tow, tug, personnel, or operational schedules. While nearly all transoceanic and coastal tows are completed uneventfully, emergency conditions must be expected. Proper preparations must be made for emergency conditions.

Good planning, preparation for emergency situations and correct ship handling are necessary elements of towing. Special towing missions generally transit unprotected coastal waters and the open ocean. These missions require considerably larger and stronger tugs using heavier and stronger towing gear to withstand the violent stresses encountered in open coastal and ocean seas.

Normally, these tugs have greater towing power (larger engines and overall heavier equipment and construction) and are equipped with towing machinery, such as single- and double-drum wire rope towing winches; tow wire guides, rollers, and pinions; cranes or winch/boom assemblies to handle the tow rigging; and a small workboat for boarding and inspecting the tow while en route. Towing gear for these missions include heavy chain bridles and pendants (anchor chain), plate shackles, retrieving wires, emergency towing bridles, towlines (hawsers), flooding alarms, pumps, and anchors.

***Towing responsibilities***

The command requesting tow of craft must provide the craft in seaworthy condition with flooding alarms, navigation lights, electrical power for alarms and lights, salvage gear (anchors and pumps), and towing gear (bridle, pendant, and retrieving wires). For suspect or deficient seaworthiness conditions, both the tow and towing command must agree on the risk of tow. The command accepting the tow must provide tug and towing gear to connect to the towed craft's towing gear.

The tug and gear must be seaworthy for the particular mission route and have the appropriate size, horsepower, and control to safely and successfully accomplish the towing mission. On accepting the tow, the towing command accepts full responsibility. Before accepting, seaworthiness must be verified. The tow should be refused if it is considered not fit for sea. Towing is accepted only after the tug's officers completed a comprehensive evaluation and survey of the tow.

***Seaworthiness***

Towing seaworthiness means suitable condition for the mission. This concerns all the various technical implications of the tow and towing vessel, including:-

- Vessel design and specifications.
- Structural condition and stability.
- Age, maintenance history, and status.
- Reinforcement requirements.
- Hull and superstructure closures.
- Adequacy of towing gear.
- Dewatering facilities.
- Chafing gear.
- Fire fighting and damage control facilities.

- Repair parts.
- Tow-boarding facilities.
- Emergency towing gear.
- Waters to be transited.
- Hazards of the route.

A certificate of seaworthiness for ocean tows must be completed. The certificate indicates the general characteristics of the tow, type of cargo, towing gear, lights, and emergency gear aboard the tow.

***Categories of risk***

Commanding officers of the towing ship and the towed craft should agree to the conditions of risk in towing the craft. Risk conditions are based on the seaworthiness and structural condition of the tow, expected sea and weather conditions for the route, and the specifications of the towing ship. In summary, foreseeable hazards are as a rule considered by the commanding officers in advance and suitable precautionary measures are taken to reduce the occurrence of accidental events during the voyage. Accidental events can still occur in spite of precautions because of the inherent risky nature of the towing operation.

A certain marginal risk cannot be completely eliminated for various reasons. The main reasons are the weather conditions, hidden faults in ship, tug and gear and not at least human errors. Such marginal risk is difficult to model in detail and it is according necessary to model it "on average" by using statistics of accidents that have occurred in the past. Statistics for oceanic towing accidents have not been available and statistics for normal ship operation have been used.

This is considered as a good approximation as in both cases the main risk contribution is the rate of occurrence of human errors. The risk estimate is given in the next sub-section.

***Methodology and basis for risk assessment***

The methodology and basis for the risk assessment is described below. The risk of occurrence of towing accidents is dominated by human errors as is the case for ship accidents in general. This is documented by statistics from U.S. Coast Guard shown in Table.

**Table. Human error contribution to various types of accident**

| Type of Accident | Human Error Contribution |
|---|---|
| Tanker accidents | 84 - 88% |
| Towing vessel groundings | 79% |
| Collisions | 89 - 96% |
| Allisions | 75% |
| Fire and explosions | 75% |

A search in published papers, (e.g. Journal of Navigation) and on the internet has indicated that available statistical models and data for ship accidents do not cover towing accidents.

As towing accident models are not available and as towing accidents are dominated by human errors as ship accidents are, accident models and data for ship accidents are used to predict the frequency of accidents under towing of ships from Europe to Asia. The generic simplified ship accident model used is as follows:

- accident rate per ship mile x ship miles

This simplistic model describes the situation in general. It has been developed by dividing the number of accidents with the number miles sailed for certain area over certain periods of time. The generic accident rates are given Table. All the mentioned types of accidents are included in the risk estimate given in Table.

**Table. Generic ship accident rates for differnt types of accidents according to DNV**

| Type of Accident | Generic accident rate per ship-mile |
|---|---|
| Collision | $5.0 \cdot 10^{-7}$ |
| Grounding | $7.2 \cdot 10^{-7}$ |
| Fire/explosion | $1.0 \cdot 10^{-8}$ |
| Hull damage | $1.6 \cdot 10^{-8}$ |

**Table. Risk per towing voyage London**

| | London - Alang via Suez Canal | | | London - Alang via Cape of Good Hope | | |
|---|---|---|---|---|---|---|
| | Probability Number | Probability Percent | Recurrence Number | Probability Number | Probability Percent | Recurrence Number |
| Factor | 10000 | 1 | 1 | 10000 | 1 | 1 |
| Accident | 124.9 | 1.249% | 80 | 215.5 | 2.155% | 46 |
| Total loss | 1.2 | 0.012% | 8006 | 2.2 | 0.022% | 4640 |
| Fatalities | 8.3 | 0.083% | 1201 | 14.4 | 0.144% | 696 |

The accident rate for collisions, for example, means that 5 accidents may be expected for every 10 million sailed nautical miles. The inherent risk in towing of vessels is in general larger than if the vessels are self-propelled because the complexity in towing is greater, i.e. there are more hazards in towing that in normal sailing.

The additional hazards may come from handling of the towing arrangements, communication between the tug and the towed vessel, more severe weather restrictions, low quality towing companies undercutting the price etc. A risk increase factor is applied here to account for the additional hazards. The risk increase factor has been evaluated by two methods:

- increased insurance premium for towing of a ship compared to when it is self-propelled;
- guideline for ship collision with bridges by both self-propelled ships and ships under tow.

The additional yearly insurance premium for towing a ship from Europe to India expresses the insurance company's assessed extra risk compared to normal self-propelled sailing. This additional yearly premium has been estimated at 8.5% by a marine insurance company.

This gives an average 60% increase for the duration of towing that has been estimated at 60 days. It has not been possible to distinguish between the two routes. The American State Highway Organisation has developed a standard and guideline for assessment of the risk of vessel collision with highway bridges. (AASHTO LFRD Standard). In the standard, the generic probability of mismanoeuvre at bridge passage is a factor 2 larger for towed barges than for ordinary ships.

The two estimates are considered in agreement for the present purpose and a risk increase factor of 1.6, equal to an increased risk of 60%, is applied on the accident probabilities to give the results in Table. Furthermore, the risk estimate is based on the following data:

- The distance London - Alang through Suez is estimated at 6265 nautical miles according to Portworld.
- The distance London - Alang around Cape is estimated at 10,810 nautical miles according to Portworld.
- An average towing speed of 5 knot is used.

## COSTS OF SHIP DISMANTLING AND PRECLEANING OF SHIPS

This section presents an analysis of the costs and benefits of ship dismantling and pre-cleaning of ships. The objective has been to derive the additional costs of pre-cleaning and dismantling of ships applying sound environmental and safety standards, thus making it possible to compare the environmental advantages with the short-term economic disadvantages of pre-cleaning and ship dismantling at "green" facilities.

The present approaches to green recycling are more expensive than the conventional beach breaking. The extra cost of performing green recycling is related to a demand for infrastructure (drainable surfaces, cranes etc.), and to each of the extra processes involved in green recycling, including:

- Construction costs for new equipment, machinery and infrastructure;
- Manpower costs for new and more time consuming work routines;
- Hazardous waste disposal costs.

Dismantling of ships is time-consuming and labour-intensive. The time used for performing cleaning of vessels between 10,000 and 25,000 LDT was found to range from 4 to 7 weeks (COWI/DG TREN, 2004). Therefore the cost of labour plays a very important role for the overall economics of ship breaking. Ship owners generally receive positive cash flow from the demolition market when selling ships for demolition.

The ship owner will typically sell the ship for demolition to the ship breaking facility associated with the highest net profit. Here the price offered by the ship breaker is most important, but transaction costs might also play a role. The total price of the vessel offered by a ship breaker is determined by a number of site specific factors such as steel prices, investment costs, labour costs, health and safety requirements, taxes and levies etc.

## APPROACH

The approach for analysing the costs and benefits of ship dismantling and precleaning of ships is described below. This includes a description of scenarios and effects included in the analysis. The analysis of costs and benefits is carried out from a ship owner perspective (EU located) considering the following scenarios:

- Reference scenario. A business as usual scenario in which the ship owner will sell the vessel for scrapping in Asia where the conventional beach breaking method will be applied.
- Pre-cleaning scenario. Pre-cleaning of ships in EU and Candidate countries and final dismantling at beaching sites in developing countries (Asia).
- Full "green" dismantling in EU scenario. Dismantling of ships at a "green" facility in EU complying with applicable EU law and therefore also with the developed recycling guidelines under a new convention.
- Full "green" dismantling in OECD scenario. Dismantling of ships at a "green" facility in an OECD country (represented by Turkey) complying fully with the developed recycling guidelines under a new convention.

Pre-cleaning of vessels destined for scrapping consists of removal of the contaminant and hazardous materials. The objective is to neutralise the vessel in such a way that its dismantling in existing capability, ideally, would not create environmental problems. This requires that the vessel to be stripped of environmentally hazardous materials such as asbestos, fuel oil, lead, PVC, PCBs and more.

A pre-cleaned ship is a hulk, a "dead" ship without power systems. The vessel will need to be transported to the scrapping facility under tow. The costs of towing will impose an expense and would have to be accounted for in the overall economic analyses. Furthermore, towing is an operation that involves an increased risk of accidents that may for example lead to loss of ship.

The scenarios analysed are illustrated in the table below and split into EU and non-EU according to the existing conditions for the reference scenario and the applicability of the EU Waste Shipment Regulation for the future scenarios.

**Table. Scenarios analysed**

| | Pre-cleaning | | Breaking | |
|---|---|---|---|---|
| | EU | non-EU | EU | non-EU |
| Reference scenario* | | √ | | √ |
| Pre-cleaning scenario | √ | | | √ |
| Full green dismantling in EU scenario | √ | | √ | |
| Full green dismantling in OECD scenario | | √ | | √ |

* Conventional beach breaking method

The table illustrates where cleaning and breaking takes place in the various scenarios. Obviously, cleaning may be performed as an activity separated from the breaking or performed in relation and coincident with the breaking of the ship.

The table shows that there are three alternatives to the reference scenario which have been analysed. The results of these scenarios are compared with the results of the reference scenario. The calculation of costs and benefits is restricted to these scenarios. On this basis the consequences of other cleaning and dismantling scenarios are discussed. Theoretically three different levels of pre-cleaning could be relevant:

- To the extent which still just allow the ship to sail by itself (self propelled);
- To the extent which allows export of the vessel under the existing EC Waste Shipment regulation, i.e. "properly emptied";
- To the extent where the ship is completely emptied of hazardous materials.

As earlier described, level 1 would not meet WSR requirements and level 3 is not a practically achievable level. In this section on economics it is level 2 of pre-cleaning that is considered in relation to analysing the costs and benefits of ship dismantling.

The time and resources needed to clean and dismantle a ship depends on the type of ship. Ideally the analysis should be carried out to reflect the consequences for all the different types of vessels. However, it is not possible to carry out detailed analysis for all types. It is presumed that a tanker is a good representative for the different ship types.

Therefore the calculation of costs and benefits is carried out for a medium sized tanker. Costs differences between different types and different sizes of ships must be expected. Based on the analyses of the consequences for a tanker it is discussed how to dismantle other types of vessels (i.e. passenger ships, naval ships etc.) which influences the results of the different scenarios respectively. The aim has been to estimate the costs of dismantling (or net profit as a vessel often represents a value for the ship owner) per LDT for each scenario thus making it possible to estimate the additional costs of safe and environmentally sound ship recycling. The table below lists the effects which have been estimated in the different scenarios.

**Table. Effects included in the analysis of total cost/benefits for ship dismantling in the scenarios**

| | Reference scenario | Pre-cleaning scenario | Full green dismantling scenarios |
|---|---|---|---|
| Transport costs for last voyage* | X | X | X |
| Risk of accident on the last voyage | X | X | X |
| Pre-cleaning costs | | X | |
| Towing costs from pre-cleaning to dismantling site | | X | |
| Risk of accident by towing | | X | |
| Ship end-of-life value ** | X | X | X |
| **Net revenue from recycling** | | | |

* 'Last voyage' meaning after a cargo has been delivered in last port of call to the scrapping/pre-cleaning facility.

** The ship end-of-life value is for the ship in the current state, which for example means the pre-cleaned ship in the pre-cleaning scenario. The value is expected to vary considerably between the different scenarios first of all as a consequence of the different recycling standards and techniques applied.

The transport costs are the costs of sailing an empty ship on its last voyage from the point of departure to the final destination, i.e. the dismantling or precleaning site. The cost from risk of accident is the expected costs of losing a ship. It is calculated as the risk of accident multiplied by the revenue from ship sale. The pre-cleaning costs have been estimated using different approaches for deriving the costs.

This estimate is based on interviews with a number of key stakeholders about the costs and actual prices, and on the present work's calculations of the pre-cleaning costs based on observed transactions. Interviews with key industry players have been carried out to estimate the costs of transportation (towing) of the vessel to developing countries.

The costs of dismantling have also been obtained using different approaches to derive the costs. First of all, a top-down approach in which the price of dismantling has been directly determined from contracted transactions (or directly by information from the industry) has been applied.

This means that the price has been based on data from the Clarkson database. The prices available from Clarkson mostly reflect the price for scrapping of vessels in Asia where the conventional beach breaking method is applied (the reference scenario). Unfortunately, it is not possible to extract information about prices for vessels that have been scrapped at a "green" recycling facility from Clarkson's as the number of facilities offering green recycling is limited in Europe and globally. Until recently, the market for such services has been limited to a few governmental programmes and concerned companies, and a competitive market for these services has yet to emerge. Green recycling of ships is, generally, considered more expensive than traditional breaking due to the costs of measures taken for the benefit of the environment and worker's health and safety. The figure below shows scrap prices per LDT for the last three years for selected Asian countries.

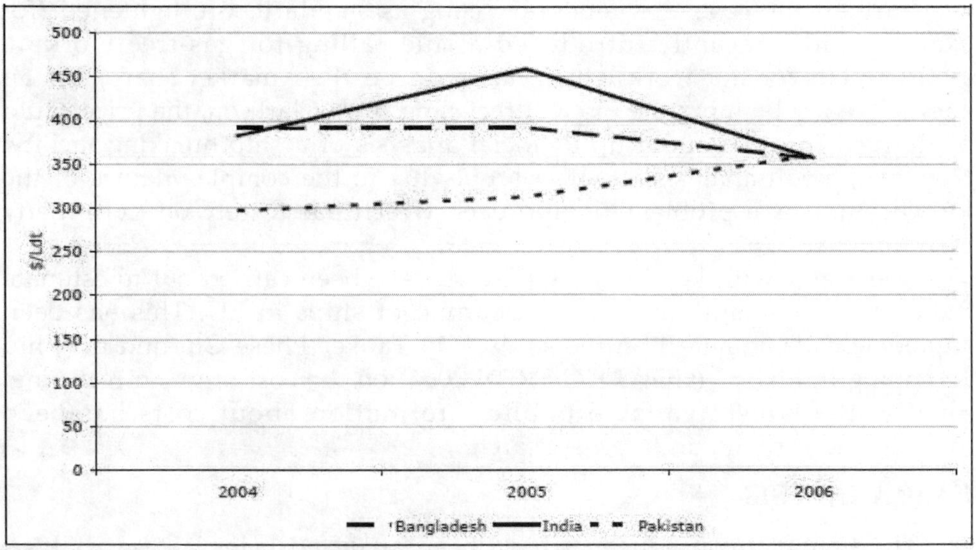


**Fig.** Scrap price by country

The figure shows that there are some variations in prices between countries. The variations are due to several factors including differences in labour costs, local steel prices, local taxes and local differences in prices for second hand materials. However, the type and the size of vessels scrapped are also important: Due to a favourable ratio of steel to non-reusable items tankers generally fetch higher prices, whereas other vessels such as passenger ships, ferries, and car transporters may command 100 USD less per tonnes.

**Table. Average scrap prices by type**

| Type | Average USD/LDT | Number scrapped |
|---|---|---|
| Oil tanker | 399 | 177 |
| Gas | 386 | 15 |
| Other cargo vessel | 354 | 38 |
| Container | 333 | 2 |
| Bulk carrier | 324 | 43 |
| Other tanker | 300 | 3 |
| Passenger/ro-ro/vehicle | 266 | 10 |

In ship breaking countries in Asia, costs associated with health, safety and environmental standards are relatively low, as these are not an issue at most recycling facilities. Hence, costs associated with health, safety and environmental issues play only a minor role compared to labour costs. However, they probably still explain a small part of the cost difference between countries in Asia.

Further, the costs are generally rising as standards are tightened (for example India recently introduced a rule calling for gas-free working environment for hot work which has reduced their market share). As an alternative to obtaining the price of recycling from Clarkson, the price could be derived from a bottom up financial analysis of a ship breaking facility. However, a full analysis is out of scope due to the complex nature of the industry and is a problematic approach which has to rely on (too) many assumptions.

Interviews with key industry players have been carried out to estimate the costs of dismantling and pre-cleaning of ships in EU. This has been supplemented by data from the study "Oil Tanker Phase Out and the Ship Scrapping Industry" (COWI/DG TREN 2004) on the cost of green recycling. Finally, the latest available public information about costs has been investigated through an internet search.

## CASE ANALYSIS

The financial consequences have been analysed using a case analysis approach. As mentioned above a tanker is a good representative for the different ship types and therefore the calculation of costs and benefits has been carried out for a medium-sized tanker.

### General assumptions

The analysis has been built up around an oil tanker which was actually scrapped in Asia in 2006. Exact characteristics of the tanker are provided in the table below.

**Table. Characteristics of the tanker selected for case analysis**

| **Name** | ***Venture III*** |
|---|---|
| Type | Oil tanker |
| DWT | 82,006 |
| LDT | 15,432 |
| Date of demolition | May 2006 |
| Country of demolition | Bangladesh |
| Price (USD per ldt) | 380 |

The table shows that the selected ship is an oil tanker that was scrapped in Bangladesh in May 2006 for a price of 380 USD per LDT. The following general assumptions have further been applied:

- Breaking and pre-cleaning destination EU: United Kingdom, London;
- Breaking destination Asia: Bangladesh, Chittagong;
- Breaking destination OECD: Turkey, Izmir;
- Towing route United Kingdom to Bangladesh via Suez Canal or round Cap of Good Hope (cheapest option).

**Costs of dismantling and pre-cleaning of ships**

Below, it is briefly described how the cost elements in the case analyses have been estimated. It should be noted that it is a quite difficult task to estimate some of the cost elements.

The estimates and calculations are associated with substantial uncertainty and as a consequence the quantitative analysis is more indicative than precise. It provides the right order of magnitude rather than the exact level.

***Transport costs for last voyage and risk of accident on the last voyage***

Conventionally, a commercial vessel will carry cargo to Asia to a port close to the breaking beach. It has therefore been pragmatically assumed that the destination for a vessel's last voyage with cargo is so close to the final recycling destination (or pre-cleaning destination) that the transport costs for the final leg and the costs from risk of accident on the last voyage are ignored (or estimated being close to 0).

***Pre-cleaning costs***

It is difficult to generalise the costs of pre-cleaning as it depends on a number of factors specific for each vessel, such as ship size, type, cargo type, maintenance level and repairs and refitting. Therefore estimates are highly uncertain. In general the costs of performing pre-cleaning comprise:

- manpower costs for safe removing of the hazardous waste,
- hazardous waste disposal costs,
- construction costs for new equipment, machinery and infrastructure necessary for safe separation and containment of hazardous materials from the ship.

In the study "Oil Tanker Phase Out and the Ship Scrapping Industry" that COWI performed for DG TREN, the cost of performing pre-cleaning of a ship and disposal of hazardous waste was estimated by industry sources to be around 25 - 50 USD/LDT in Asia, but higher in Europe due to higher labour costs. In this study we have tried to investigate prices of actual pre-cleaning of ships in EU.

- The *Sandrien* is a chemical tanker of 8,380 LDT built in 1974. The ship was owned by an Italian shipping company when it arrived in

Amsterdam in August 2000. Shortly after its arrival in Amsterdam harbour, the ship was detained by the port authorities due to its technical state. The ship inspectors decided that the vessel was only allowed to leave Amsterdam after undergoing fundamental repairs. The inspectors further informed the environmental inspectors on the suspicion that the ship contained hazardous materials and could be bound for scrapping. In February 2001, the Dutch Environmental Inspectorate arrested the *Sandrien* as the ship contained hazardous materials, for instance asbestos and heavy metals, and there was a contract on scrapping the ship in Asia. Such export to Asia would violate the European Waste Shipment Regulation. After several court procedures the Council of State in the Netherlands in June 2002 ruled that an end-oflife vessel not properly cleaned of hazardous materials should be classified as hazardous waste. Consequently *Sandrien* was detained in the Netherlands for cleaning and scrapping. The Dutch Ministry of Environment (VROM) is a key stakeholder in the affair. According to the Ministry (Mr. Cees Luttikhuizen) the cost of removing the asbestos from the ship accounted to approx. 700,000-800,000 EUR and a fair estimate of the costs of removing other chemicals and waste is approx. 100,000 EUR. Hence the total cost of "pre-"cleaning of *Sandrien* amounted to app. 800,000-900,000 EUR. This is equal to approx. 100 EUR per LDT.

- The report "National Guidance: Best Management Practices for Preparing Vessels Intended to Create Artificial Reefs" by U.S. Environmental Protec tion Agency (2006) provide guidance recommending environmental best management practices to be used in the preparation of vessels for use as artificial reefs. Even though the cleaning is not done according to the same standards applied to pre-cleaning for scrapping, the process is comparable.
- For the small vessel *USS Spiegel Grove* the cost per LDT accounted to approx. 70 EUR. For the larger carrier *Oriskany* the cost per LDT accounted to approx. 190 EUR.
- The pre-cleaning cost of *HMS Scylla*: 1.2 million £ for 2,500 displacement tonnes. At the present rate of 1.94 USD/GBP this is approx. 2.3 million USD or 930 USD/LDT. It should be mentioned that this figure also includes demilitarisation costs.
- Pacific Nuclear Transport Limited (PNTL), a subsidiary of British Nuclear Fuels (BNFL), exported two of its decommissioned nuclear fuel carriers to the Netherlands for recycling. The owners could have netted approximately one million pounds per vessel if they had been recycled in India due to the value of scrap steel at the time. However, recycling of the vessels in the Netherlands resulted in a cost to PNTL

of several hundred thousand pounds per 4000 tonne vessel (UK Defra 2006).

Though highly uncertain, the price of pre-cleaning of the *Sandrien* is assessed to be the best proxy for the price for *Venture III*. Taking account also of the information about cleaning of the American vessels it is estimated that the precleaning costs will be in the range of 75-150 EUR per LDT (100-200 USD per LDT).

***Towage costs for towage from pre-cleaning to dismantling site***

The towage costs have been estimated based on information obtained from interviews with key stakeholders involved in the towing business.

- Towing of the ship *Otapan* from Amsterdam to Turkey. The Dutch Ministry of Environment (VROM) was involved in the planning, which was undertaken by Landfall. The total price was 205,000 EUR, which reflected that Landfall would return with another ship. However, Otapan had to be returned to Amsterdam as the deal with the Turkish scrapping company was cancelled. The cost of the return-tow was 350,000 EUR and the operation took 30 days. On this basis a price of towing a pre-cleaned ship can be estimated to app. 8,000-10,000 EUR per day. This estimate was confirmed by Joep Caesmaat from Landfall.

The price of towing around Cape of Good Hope is calculated using an estimated cost of 10,000 EUR per day and estimated 85 days (travel speed of 5 knots and a distance of 10,800 nautical miles); so the total costs of towing of *Venture III* from London to Bangladesh would have been 850,000 EUR or 55 EUR per LDT (65 USD per LDT).

The costs of passing through the Suez Canal can be obtained from Leth Agencies. From their homepage the total toll for Suez Canal transit can be calculated for a specific ship. The toll for *Venture III* has been calculated to approx. 295,000 USD.

Again using an estimated cost of 10,000 EUR per day and a duration of 55 days, the costs of towing of *Venture III* from London to Bangladesh is estimated to 550,000 EUR. The total price including Suez Canal toll is thus 845,000 or 55 EUR per LDT (65 USD per LDT). This confirms the common understanding in the industry that the price is set so that it is marginally cheaper passing through the Suez Canal than going around Cape of Good Hope.

***Risk of accident by towing***

The risk of accident by towing was addressed in section 6. Here it was estimated that the risk of accident by towing is only slightly higher than the risk for normal overseas navigation. The risk of losing a ship that is being towed from London to Bangladesh was calculated to approx. 0.02%. For

*Venture III* this is equal to expected costs of 120,000 EUR in total or approx. 10 EUR per LDT (15 USD per LDT).

***Revenue/costs of a ship sold for recycling (end-of-life value)***

The economics of the ship recycling industry and the forces behind demand and supply on the ship recycling market are interrelated with three other markets that ship owners operate on: the new building market, the second-hand market and the freight market.

To some extent the ship recycling market serves as a buffer balancing demand and supply in the freight market with increase scrapping when the global demand for sea transport moderates. The economics of the ship recycling industry and the forces behind demand and supply on the ship recycling market are interrelated with three other markets that ship owners operate on: the new building market, the second-hand market and the freight market.

To some extent the ship recycling market serves as a buffer balancing demand and supply in the freight market with increase scrapping when the global demand for sea transport moderates. If the price obtainable in the demolition market is a significant driver for the decision of the ship owners on when to scrap, there should be a close and positive correlation between prices and volumes of decommissioning.

However, prices and volumes are actually negatively correlated, i.e. when volumes go down, prices increase. This indicates that other factors (e.g. the developments in the freight market) are more important for the ship owner's decision.

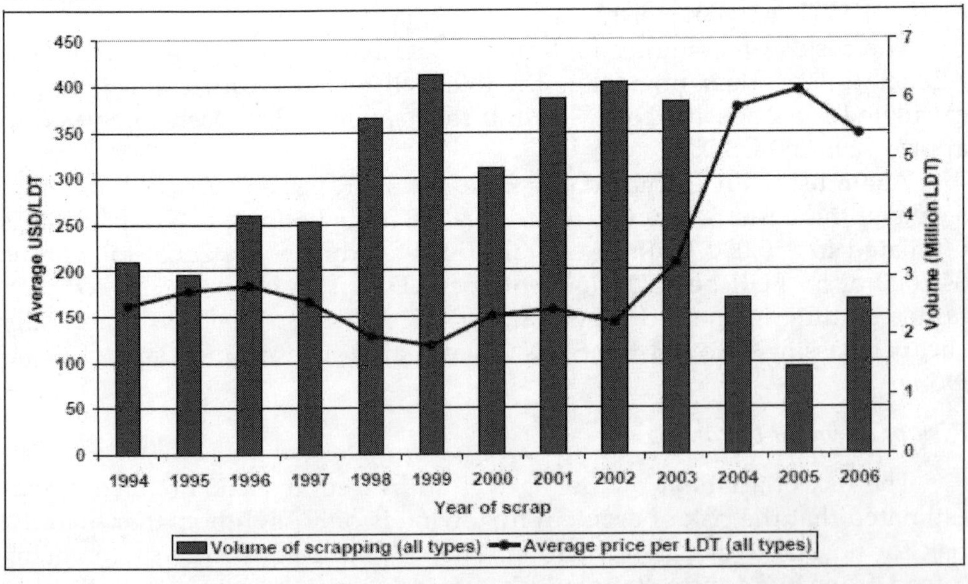


**Fig.** Average scrap price and volumes of scrap by year of scrapping (all types)

The main driver for the ship owner's decision where to scrap is the price offered by the ship breakers. With the current practice used, ship breaking is a very labour-intensive industry.

Labour costs therefore play a predominant role in determining where ships are scrapped and have been scrapped historically. The Indian Sub-Continent (India, Bangladesh, Pakistan) and China account for almost 90% of the ship breaking. This finding is generally in line with the findings of the COWI/TREN study.

There have been considerable variations in the market shares of the major ship breaking nations over the years. The figure below shows that Bangladesh today accounts for the largest share, while only 5 years back in time India was the world's largest ship breaking nation.

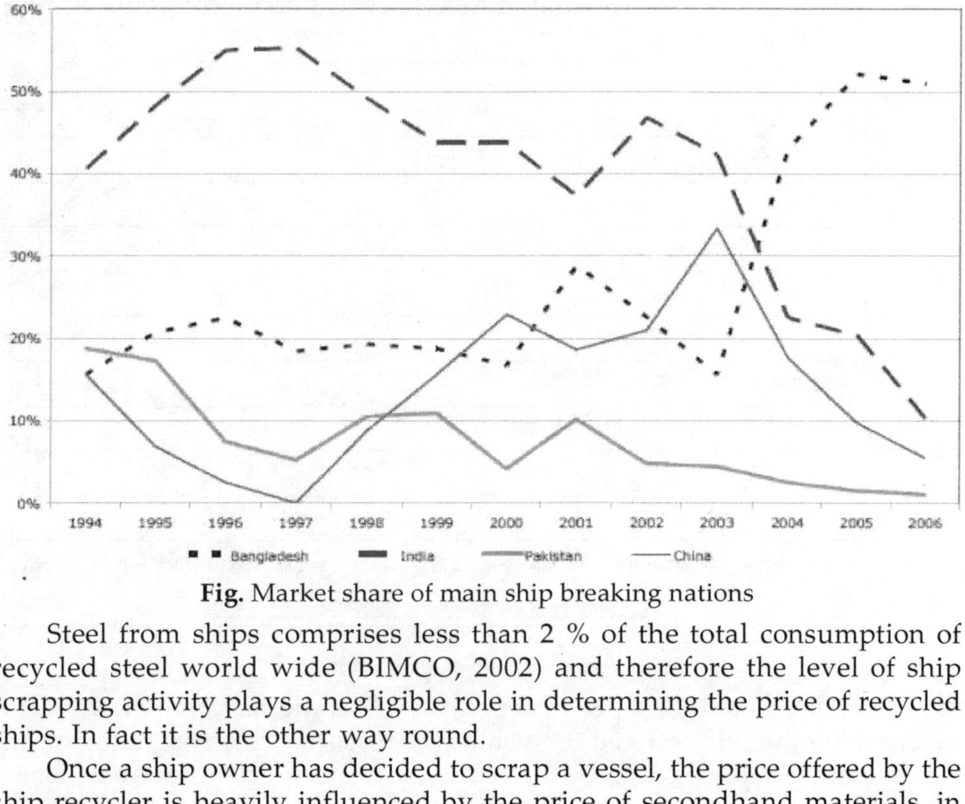


**Fig.** Market share of main ship breaking nations

Steel from ships comprises less than 2 % of the total consumption of recycled steel world wide (BIMCO, 2002) and therefore the level of ship scrapping activity plays a negligible role in determining the price of recycled ships. In fact it is the other way round.

Once a ship owner has decided to scrap a vessel, the price offered by the ship recycler is heavily influenced by the price of secondhand materials, in particular the *price for reusable steel*. When the demand for steel and other reusable items grows, steel prices increase and the ship scrappers' earning potential also increases.

As a consequence the ship scrapper's willingness to pay for a vessel for decommissioning grows (and vice versa for a weakening in the demand for steel and other reusable items). Ship breaking is a labour-intensive industry and therefore labour costs play a predominant role in determining *where* a ship is scrapped. Labour rates are 20- 100 times higher in Europe compared to India/Bangladesh.

This is by far the main reason why breaking historically has moved to Asian countries and why European countries are not operating in the market today. This nature of the recycling market means that the price of ships for recycling varies considerably over time. Below scrapping prices have been estimated as an indication of the global steel market situation at the time when the oil tanker was sold.

However, it should also be noted that the ship scrappers are operating in a global market, which is confirmed by the fact that the price trend in the scrapping countries of Asia is remarkably similar with only minor differences between countries. This is in particular confirmed by the figure below showing the average scrap prices over time for selected countries (reproduced from COWI/DG TREN, 2004).

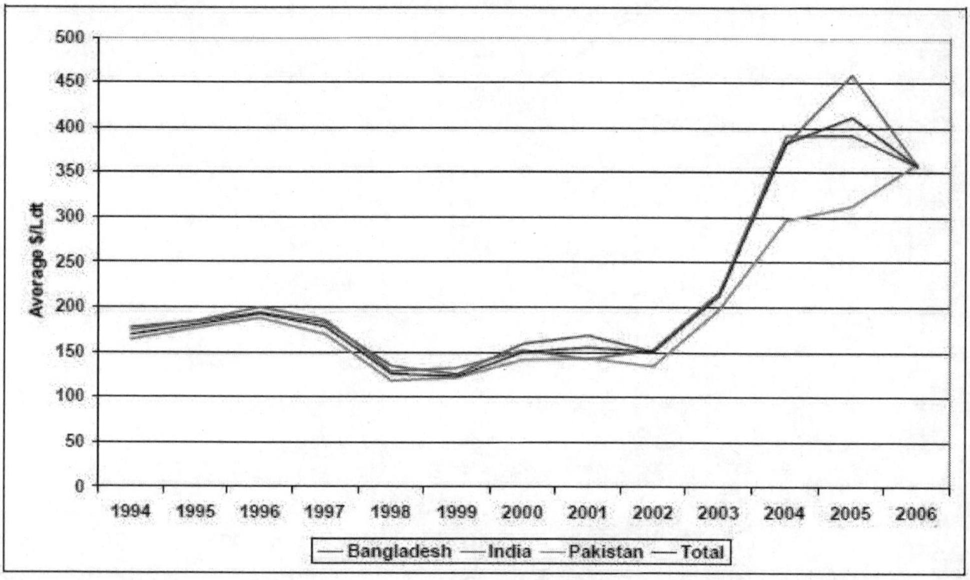


**Fig.** Scrap price—Selected countries

As earlier noted the price of recycled steel is the most important factor determining the price offered for a ship for recycling. This potentially means that the highest price could be offered by scrappers located in areas where the price of steel is highest.

However, this holds true only to a limited extent as steel is an internationally traded commodity that can be relatively easily transported from a scrapping location to a destination where it is needed. In fact, it is a common misconception that scrapping takes place in Asian countries (India, Pakistan, Bangladesh and China) only due to the fact that these countries have either a booming economy with a high construction activity or very few domestic metal resources and consequently a high demand for (recycled) steel. The cost of transporting steel from Europe to Asia is in the range of 10-15 USD per tonne, while the cost of a transport from India to China is in the range of 3- 5 USD per tonne.

These figures should be compared to prices of steel in the range of 400-650 USD per tonne depending on type. This proves that local variations in steel prices only play a small role in determining where a ship is recycled. Indeed other factors are much more important, primarily local labour costs - as recycling (using the conventional beaching method) is a labour-intensive industry - and secondly environmental, health and safety regulations.

The main drivers in the scrapping industry can simplistically be described as follows: Freight rates determine *when* to scrap. Labour costs determine *where* to scrap.

Steel prices determine the size of the ship owner's *profit*. The price of a ship that has been sold for demolition using the conventional beaching method can (most often) be directly obtained from Clarkson for most vessels. Hence, it is reported that *Venture III* was sold for scrap for a price of 380 USD per LDT, which equals approx. 5.9 mill. USD.

This reflects the price for scrapping in Bangladesh May 2006. This price level is confirmed by prices reported by GMS in May 2006 for scrapping in Bangladesh, India and Pakistan, but in some cases prices up to 500 USD/LDT have been paid. Only a few recycling facilities exist today in the EU.

They are able to perform "green recycling" of ships, but the capacity is limited and they do not dismantle the large-size ships, such as VLCC or ULCC. Hence, only few ships have been scrapped at a "green" recycling facility in the EU. As a consequence, prices paid for ships scrapped at a green recycling facility can not yet be observed in the market.

European ship recycling companies have been contacted and inquired about the costs of green recycling. However, with few exceptions the industry is reluctant to provide information about the price that they are willing to pay for a specific ship or about costs involved in the recycling process. Van Heyghen Recycling in Belgium is one of the largest active recycling companies in Europe.

Van Heyghen Recycling claimed that the company would be willing to pay approx. 100 EUR per LDT excluding, however, the costs of cleaning the ship of hazardous waste. In particular the manager emphasized that the cost of removing asbestos often proves important. The 100 EUR per LDT (130 USD per LDT) expresses the net value of the recyclable steel without labour and capital costs of dismantling the ship.

Even when the process of recycling is undertaken in EU with a higher degree of automation and use of machinery it is still a labour-intensive process. This is the main reason why the value of an end-of-life ship is much lower for a European ship recycling company than for an Asian company.

HKS Metal's facility in Gravendeel (slipway and pier breaking) mostly scraps inland-water vessels, and can dismantle vessels up to 4,000 LDT. They report that they buy vessels at 140-150 EUR/LDT (180-200 USD per LDT). This is approx. 70-80 EUR less than the steel smelters buying price to cover the costs (and profits) of dismantling the vessel.

In the price the cleaning costs for oil, fluids and asbestos are not included. These costs are usually covered by the seller of the vessel. Hence, HKS Metal's facility seems to be able to pay a higher net price than Van Heyghen Recycling. The cleaning costs in relation to a full green recycling are likely to be the same or less than those occurring when a ship is pre-cleaned because the cleaning process is easier as it can be done in an integrated process with the dismantling of the ship.

Using the pre-cleaning costs calculated above, the cleaning costs in relation to full green dismantling are estimated at 50-150 USD per LDT. These costs will heavily depend on the quantities of asbestos, lead and other hazardous waste. Hence, using these estimates the net price that a European recycling facility should be prepared to pay for a vessel ranges between 20 and 130 USD per LDT. GMS is one of the largest cash buyers of vessels for recycling in the world.

According to GMS, facilities in India, Bangladesh and Pakistan offered prices for tankers around 360-380 USD per LDT in May 2006, whereas Turkey was paying in the region of 225 USD per LDT for tankers at the same time and this was comparable to the prices in China. The price variation reflects the difference in method applied and not least the difference in labour rates.

## Case analysis results

The table below summarises the estimated effects in the four scenarios. The effects were discussed and estimated in the section above and apply to the representative case of a medium-sized tanker. Again it should be noted that estimates and calculations are associated with substantial uncertainty.

As a consequence, the quantitative analysis provides the right order of magnitude as the difference in the net profit for ship owners in the four scenarios.

**Table. Estimated costs of effects of ship dismantling in the four scenarios**

| USD per LDT | Reference scenario | Pre-cleaning scenario | Full green dis-mantling, EU | Full green dis-mantling, Turkey |
|---|---|---|---|---|
| Pre-cleaning costs | 0 | -200 - -100 | 0 | 0 |
| Towing costs from pre-cleaning to dismantling site | 0 | -65 | 0 | 0 |
| Risk of accident by towing | 0 | -15 | 0 | 0 |
| Ship end-of-life value | 380 | 380 | -20 - 130 | 225 |
| **Net revenue from recycling** | **380** | **100 - 200** | **-20 - 130** | **225** |
| **Difference to reference scenario** | **-** | **180 - 280** | **250 - 400** | **155** |

The table shows that the reference scenario results in expected revenue of 380 USD per LDT which is equal to the price paid by the cash buyer on delivery for beaching and breaking in Asia. The Turkish scrapping industry is also in a position to pay for the vessels and currently acquire vessels for demolition in the free market.

Consequently, in the scenarios with full green dismantling in Turkey the ship owner's expected revenue equals the price paid of 225 USD per LDT. The net difference to the reference scenario is approx. 150 USD per LDT. The ship end-of-life value for a European demolition company has been prudently estimated to 20 to 130 USD per LDT.

The cost difference to Turkey is primarily due to higher wages in EU countries compared to Turkey. The industry in Turkey thus enjoys an unrivalled advantage over any existing alternative European site. The minimum wage in Turkey is a fraction of the wage in many EU countries, but comparable to the minimum wages in some of the recent Eastern European members to the EU.

The net difference to the reference scenario is approx. 250-400 USD per LDT. The calculated relative cost difference between reference scenario and the full green dismantling scenario is associated with uncertainty as no ships are actually scrapped in Europe to allow direct comparison of prices paid. However, the relative cost difference has been checked using an alternative rough approach.

In a recent report by the French Secrétariat Général de la Mer the productivity in terms of steel cut has been estimated for Asian and European scrappers. It is estimated that the productivity in Europe is approximately 10 times higher compared to the Asian scrapping facilities, i.e. steel cut per worker per year is 300 to 1200 tonnes, while in Asia it is 30 to 80 tonnes per worker per year (Annexe XVI, p.143).

Taking account of the much higher labour costs in Europe, i.e. approx. 50 times higher, the production costs not taking account of capital costs are roughly 5 times higher in Europe compared to Asia.

A rough estimate of production costs could be in the order of 50 USD per LDT (steel price of 500 per tonne, price paid per LDT of 400 USD and profit of 50 USD). Using the calculated productivity factor of 5, the production costs in Europe can be estimated at 250 USD per LDT. Adding capital costs and taking account of slightly lower prices for reusable steel in Europe, a costs difference of 250 to 400 USD per LDT seems plausible.

The price disadvantage for both the Turkey and the EU full green dismantling scenario is counterbalanced by other factors that are in favour of these countries in terms of labour, safety and indeed environmental conditions. Finally, the pre-cleaning scenario has been estimated to result in expected revenue of 100-200 USD per LDT for the ship owner.

The estimate is associated with uncertainty as especially the costs of pre-cleaning have been difficult to estimate and heavily depend of the specific characteristics of the vessel, i.e. the configuration of the materials onboard a vessel. The net difference to the reference scenario is app. 180-280 USD per LDT. The calculated cost difference of the various scenarios relative to the reference scenario indicates the 'incentive gap' for the industry to choose these options over the traditional beaching in Asia.

The calculated cost difference thus essentially expresses the size of a subsidy needed in order for a ship owner to find it attractive to use a European facility over an Asian facility. The cost difference relative to the reference scenario for all alternative scenarios is substantial, which points to the fact that the options are dependent on substantial government subsidies and public funds or other regulatory measures.

## GENERALISATION AND SUM-UP

The section above discussed and estimated the effects for the case of a mediumsized tanker. Costs will differ for other types and different sizes of ships. Generally the cost difference must be expected to increase (in relative terms) with the size of ships as the man-hours needed to dismantle the ship increase with the size.

Since labour cost is the most important cost driver of dismantling vessels, this means that the advantage of low-cost countries over high-cost countries increases in relation to the size of the ship. For small ships (<2,500 DWT) the advantages of the low-cost countries is relatively smaller, which also explains why ships operating solely in Europe are often dismantled in Europe (especially in Turkey).

The costs of transporting the ship to Asia simply outweigh the relative price difference for small end-of-life ships in Asia and Europe. The ship type is also important for the price difference. Two factors play an important role: The complexity of the ship structure and the configuration of materials of a vessel. The more complex a structure of a vessel, the more man-hours are needed to dismantle the ship.

This means that tankers, bulk carriers and container ships are easier to dismantle than passenger/ro-ro ferries and non-cargo vessels. Consequently again the costs difference must be expected to be relatively higher for ships with a complex structure. The configuration of materials is also important for the man-hours needed to dismantle the ships - especially for environmentally safe dismantling in Europe. The more hazardous materials on a vessel, the more man-hours are needed.

Furthermore, the cost of disposing the waste increases with the amount. Both factors eventually lead to higher costs for cleaning the vessel. From the results of the case analysis in section 7.3 it can be noted that from a strictly economic perspective the most attractive alternative to beaching in Asia is dismantling at a Turkish site. Turkey is a member of OECD, is a Party to the Basel Convention and is a member of the International Maritime Organization (IMO).

While the solution might be economically sound/acceptable it is still to be assessed whether it is environmentally, socially and consequently politically acceptable. Economically, the second-best alternative seems to be the option with precleaning in Europe and dismantling in Asia. As discussed in previous sections of this report, there are however a number of problems and open questions associated with this option. These include:

- Environmental risk involved with towage of the vessel,
- Practical problems with the cleaning process (continued health and safety hazards for scrapping personnel),
- Practical problems of beaching the vessel under tow,
- The dangerous and hazardous working conditions embedded in the beaching method used in dismantling in Asia would remain unchanged.

On the other hand pre-cleaning has a clear advantage as it will not require any major development of facilities since it could be undertaken at existing repair yards in Europe.

The high cost of European labour coupled with the cost of complying with all workers' health and safety and associated environmental regulations in Europe is the reason why the scenario with full green dismantling in EU is the least attractive alternative. Today the capacity for dismantling all European owned/flagged vessels does not exist. There are no technical constraints for establishing the capacity.

If the framework conditions are changed so that there is a real incentive to scrap in Europe, the industry will establish sufficient capacity to meet the demand. However, it should be underlined that the EU scenario is far from being self-supporting and a scheme which ensures that there is a real incentive will have to rely on generous subsidies or other radical measures.

## DIFFERENT OPTIONS FOR DEVELOPING STRATEGIES FOR SHIP DISMANTLING IN THE EU

### BARRIERS, OPTIONS AND INCENTIVES

As pointed out in the COWI/DG TREN study (2004), the challenge for the EU - and the Member States - for developing viable strategies for ship dismantling in the EU is associated with the lack of economic incentive to perform recycling in Europe and to the global nature of the shipping business, operating without barriers to the trade of near-end-of-life ships. To avoid the scrapping of ships under unacceptable conditions a number of actions can be envisaged. Basically, there are three types of instruments available for handling this situation:

- Regulation. Although IMO is underway with a convention on ship recycling, the process is long and enforcement problems must be anticipated. EU interim measures may be necessary.

- Economic instruments. A number of instruments allowing ship owners to choose acceptable scrapping have been brought forward.
- Information. Raising awareness among ship owners, authorities in breaker countries, workers and other stakeholders.

Recommendations for these instruments should be directed towards establishing incentives for the development and promotion of green recycling capacity and to the supply of vessels to recycling in a safe manner. Regardless of the instruments chosen the core questions are:

- How do we get more green recycling capacity?
- How do we get more ship owners to use it?
- How can we provide more incentive? and
- Who is going to pay for all that?

To take the last question first, the established waste policy of the EU includes "the polluter pays principle". Also, concepts aimed at enhancing the responsible approach of the producers of services and product, e.g. product stewardship and sustainable business development, are firmly rooted in EU environmental policy.

Therefore, incentives to promote these attitudes amongst the European industrial leaders are already available although not directed specifically towards the shipping industry. In addressing the unique and strongly global nature of shipping it is possible that the industry requires its own instruments or modifications of existing instruments special for the shipping and ship recycling industry. A number of options for promoting safe and environmentally sound ship recycling have been mentioned in previous studies, among them:

- Support to the green facilities to increase their competitiveness, e.g. R&D, technical assistance, procurement of equipment.
- Subsidies, tax benefits etc. to ship owners using green recycling.
- Technical assistance to the development and implementation of national roadmaps to improved conditions in breaker yards.
- Assistances to the shipping industry, e.g. with development of certification/ auditing activities for green recycling facilities and lists of approved facilities.

Considering the development over the last 30 years in European and global ship recycling, it is obvious that today's ship recycling takes place where the costs are the lowest.

In the shipping industry where the assets, in this case the hulls, are movable with few regulatory strings attached, there is no economic incentive to choose a European yard regardless of its compliance with any safety, health and environmental regulation applicable. The one force that closed the local European recycling industry and fostered the Asian was profitability. More than anything, the profitability is based on the cheap labour costs in the beaching nations of Asia, but also the ignorance of basic safety, health and environmental considerations and the almost complete absence of investment needs for infrastructure, and adds to the economic lure of beaching. There is no advanced technology associated with ship recycling and safe recycling of ships can take place anywhere in Europe or in upgraded "green" facilities elsewhere in the world.

However, the willingness of the recyclers to invest in facilities delivering "green" recycling is based on their expectations for the market. Today, one of the primary barriers to the economic feasibility of a "green" yard is how to ensure a steady supply of vessels to the facility, if the not so eco-friendly yard next door or in a neighbouring country offers to pay 100 USD more per LDT?

## ENSURING A SUPPLY OF VESSELS

### Government vessels

One of the ways to provide raw material to the "green" yards is by channelling government-owned vessels to the yards. Supply from governments in terms of naval vessels is 40-45,000 LDT/year. This corresponds to one or two yards the size of the existing European recyclers of inland and smaller vessels.

It must be recalled that not all of these vessels go to Asia or Turkey; with the possible exception of the largest (frigates, destroyers and air craft carriers) many of them are recycled at the existing yards in Europe. The large vessels are the challenge. Dismantling a large vessel obviously requires more space, may require bigger cranes and other investments. If the extra supply of government vessels falling under the WSR is to be leverage for a recycling capacity in the EU that may also address the commercial market, public international tendering of these government vessels should be enforced.

Otherwise there will be a risk that the most efficient recycling yard(s) cannot access their potential supply base. It is assumed that naval vessels and other government vessels will be recycled in the EU or OECD, thus not violating the WSR, and no need for additional regulation or incentives are needed with the exception of public tendering.

### Merchant fleet

The tonnage of merchant vessels under European flags are a much larger volume and could form a much more stable supply base for a European recycling industry.

A considerable tonnage is in theory affected by the WSR (up to some 2.1 million LDT/year), but in reality few vessels are directly affected by the Regulation since ships flying the flags of the EU Member States may very well become waste outside of Europe.

Also, there are no legal obstacles to the transfer of an old vessel to an other flag or it may be sold and resold and soon end as a scrap vessel. In some flag states the implementation and enforcement of the law regarding the scrapping issues may be expected to be somewhat lax.

However, at some point the competent European authorities could choose to consider a certain reflagging rate and sale volume of former European-flagged vessels as a *de facto* evasion of the WSR leading to export to the existing substandard facilities.

It has been suggested from NGOs and some member states that not the flag state, but the country of the company owning or controlling the vessel should determine the applicability of the WSR. It is beyond the scope of the present study to establish the potential consequences for the European shipping industry including the future company registration pattern when it comes to ownership of near end-of-life vessels. For the purpose of the study the recycling of 1.3 - 2.1 million LDT/year presently operating under EU member state flags are sufficient to demonstrate the range of costs associated with the establishment of a "green" recycling capacity.

This number may be somewhat inflated as the trading pattern for EU member state flags often involves selling older vessels for continued trading and investing in a younger fleet. At the projected time of scrapping a significant part of the vessels may no longer be flying the colours of a European member state.

A limited number of European ship owners already developed company policies on ship recycling that oblige them to safe recycling of their own vessels. These include P&O Nedlloyd (now Maersk Line), Shell and BP Shipping. It must be expected that these company are prepared to pay the premium needed to finance the EHS investments, i.e. the companies are willing to cover the difference between beaching (at some 350-400 USD/LDT) and the price level at acceptable yards.

Although not explicitly stated, the companies generally maintain that the upgraded procedures in Turkey and China are acceptable as safe and environmentally sound ship recycling, and price offered from these facilities amount to some 220 USD/LDT.

The responsible ship-owners therefore experience a loss of 130 USD/LDT compared to their less conscious competitors.

The possibilities for providing incentives for the supply of vessels from the merchant fleet include:

- Rigorous legislation on sale conditions (e.g. "vessel must be scrapped according to WSR and EU standards regardless of future flag"): *This will obviously cost the ship-owners when selling. Enforcement should be via the arbitration courts, but it is not hard to foresee that recovery of profits in violation of contract conditions 5-10 years after a sale will be very difficult.*

- Take-back approach: *Developing a "take-back" approach where the building yards are charged with the responsibility to have the vessels demolished has virtues in terms of forcing product sustainability thinking onto the manufacturers, but the 25 – 30 years time lag is problematic. A global fund has been proposed and is a possibility although only a truly global levy system will ensure coverage. A problem is that it may have to be the owner of today that ends up paying for the years of profitable usage by other owners.*
- Full or partial coverage of the loss associated with "green" recycling compared to beaching.: *The approx. 130 USD/LDT less in Turkey or China could be subsidised upon presentation of a Completion form or deductible on the tax.*

## STRATEGIES TO STRENGTHEN CAPACITY IN EU BASED FACILITIES

It is a challenge to ensure the use of the presently available capacity of European green recycling facilities and expand the capacity in Europe, when the supply base is limited to government and fishing vessels. The shipping industry and the merchant fleet are optimised to deliver efficient transport services from port to port, and the current law is directed to facilitate this for a truly globalised industry.

The end-of-life situation for vessels has not until the last decade come under scrutiny, and since the activity is land-based, this has been considered a matter of national authorities and no responsibility is placed on the ship owner in that respect.

The European Waste Shipment Regulation applies to the scrapping of vessels and floating structures, but it has been very difficult to prove a ship owner's intent to dispose of a vessel while under EU jurisdiction. Only ship owners with a very clear company policy will scrap ocean-going vessels in Europe today, since the net loss will be several million Euros compared to a beaching in Asia.

The development of a convention on ship recycling in IMO will assist in transforming the shipping and recycling industries. However, if recycling countries are slow in ratifying or the flag states lack in enforcing or some countries refrain from becoming a Party in order to develop their own low-standard scrapping industry, the impact of the convention may be limited.

Anyhow, even if the convention holds the promise of "equivalent level of control", the ratification process means that it will not enter into force until at the earliest in 2013. Interim measures will be needed both to ensure the phase-out of unacceptable practises and to assist in providing green recycling capacity to the countries that adopt the new rules already upon signing the convention, incl. the EU Member States.

Several companies in the EU are developing capacity for green recycling, but none sees a profitable business in direct competition with Asian ship breaking yards.

There are only limited options for enforcing scrapping in the EU if ships are reflagged before scrapping and it may be argued that green recycling of a vessel can be carried out equally good in Asia and in the EU, if properly developed.

Even if Asia is accepted as the present and future prime ship scrapping destination, problems remain in developing technical solutions to the extent of the necessary pre-cleaning of ships, the discontinued use of the beaching method and other requirements of a coming convention.

**Development support to green facilities in the EU**

Various instruments besides "command and control" can be brought into use to either support the European or the developing countries in green recycling. These may in Europe include support from EU R&D programmes, regional and SME (Small and Medium-sized Enterprises) development programmes or direct subsidies.

There are several green recycling initiatives in EU Member States that can provide good practice working methods that could be promoted at international level and/or exported to developing countries. Those initiatives would prove that environmentally sound recycling practices are compatible with an economically viable recycling activity.

The Dutch facility Ecodock has been under planning for some years now. It was expected to be on the market in 2006, but has yet to move into the physical world. The difficulties experienced show that even with support of several EU programmes and private investors, the market conditions are not favourable for green recycling.

Increased support to development of green facilities in EU and to the improvement of Asian breaker yards probably could deliver "fast-track" solutions by upgrading existing facilities with demonstration projects, but they should be coupled to improve the market conditions.

**Subsidies**

The largely unregulated existing ship breaking capacity in Asia has for some time now been in the spotlight for not adhering to even basic concerns for health, safety and environment.

The market is very competitive on price and the industry has historically been seen to respond to increasing costs and legislative demands by moving the clusters of breakers to countries of low cost, higher demand for recycled steel and a lax enforcement of regulation. This means that under the current market conditions there are no incentives to EU ship owners to dismantle ships at a green facility.

Unless instruments are developed and/or regulations are implemented and enforced to avoid this, the incentive will be to continue the current practices. While in normal situations, EU subsidy of industry should be discouraged, it is vital that the EU find mechanisms and incentives to spark the industry.

Therefore the subsidy instrument is treated in this section even though it might require amendments to the EU guidelines for environmental state aid as they currently do not include ship dismantling or pre-cleaning as eligible for state aid. Also not considering the question of equity, i.e. who should bear the costs, using subsidies is the simplest way of making green recycling economically attractive.

Moreover, subsidies can be used as a means to provide a sufficient incentive for ship owners to sell their ships to recycling facilities complying with the guidelines.

The additional cost for the ship owner of three alternatives to conventional beaching was calculated. In the cheapest alternative, dismantling in Turkey, the ship owners' expected revenue is reduced by approx. 150 USD per LDT.

The pre-cleaning scenario is expected to result in additional costs of 180- 280 USD per LDT, while the scenario with full green dismantling results in additional costs of 250-400 USD per LDT. In future scrapping volumes were estimated. Scrapping of vessels which today are both EU-flagged and EU-owned will amount to 1.1-1.8 million LDT/year, except for 2010, and EU-flagged vessels which are owned by non- EU countries will account for scrapping of 0.2-0.3 million LDT/year.

On average from 2007-2020, EU-flagged scrapping will amount to 1.6 million LDT/year. On this basis the theoretical level of subsidy needed to ensure that environmentally acceptable dismantling is competitive in relation to conventional beaching in Asia can be calculated.

**Table. Estimated subsidy needed for ship dismantling in the three alternative scenarios until entry in to force of new convention**

| | Pre-cleaning scenario | Full green dismantling, EU | Full green dismantling, Turkey |
|---|---|---|---|
| Cost difference to reference scenario (USD per LDT) | 180 - 280 | 250 – 400 | 155 |
| Scrapping volume, all vessel types, EU flagged (LDT per year) | 1,600,000 | | |
| Total subsidy needed, 2007-2013<br>mill. USD per year<br>mill. EUR per year | <br>290-450<br>214-333 | <br>400-640<br>296-473 | <br>250<br>185 |

China is not shown in the table, but it is estimated to resemble the case for Turkey. The table above presents the total cost difference between continued uses of conventional beaching for EU-flagged ships and the alternative approaches analysed. It highlights that in total subsidies of approx. 250 million USD per year would be needed to provide sufficient incentive for ship owner to choose Turkish sites over Asian.

A larger subsidy in the range of 290-450 m. USD per year would be needed to support the pre-cleaning alternative, whereas 400-640 m. per year would be needed to support full green dismantling in EU.

A subsidy scheme for "green" recycling should not apply indefinitely. Assuming that the proposed IMO draft Convention is approved and signed by the EU member states and possibly the EU as a Party, the entry into force of the Convention on Ship Recycling may at the earliest take place in 2013.

At that time a level playing field will be established and no subsidies should be required. The promotion of and support for green recycling should be limited to the interim period up to the expected entry into force. The subsidies could potentially take a number of different routes, e.g.:

- Support to the green facilities to facilitate capacity-building and increase their competitiveness, e.g. R&D, technical assistance, procurement of equipment.
- Direct subsidies to ship owners using green recycling.
- Direct subsidies to ship breakers to make them competitive.
- Tax benefits etc. to ship owners using green recycling.

The subsidy could be paid directly to ship owners choosing an approved green recycling facility. The subsidy needs to be of a size which makes the total benefit (subsidy plus price paid by the ship scrapper) of choosing a green recycling facility as high as choosing an existing non-complying facility.

Designing an efficient scheme with subsidies to ensure green dismantling of EU-flagged ships is associated with many challenges and potential pitfalls.

At the same time, the subsidy should be sufficient to make green dismantling competitive, yet it should not be so high that green facilities in EU/Turkey would become more attractive for the world's ship owners. If the subsidy is too low dismantling in Asia will continue to prevail.

On the other hand, if the subsidy is too high it could lead to an adverse problem of the world fleet being re-flagged to EU flags to enjoy high end-of-life ship values for scrapping at a green facility. Deciding the appropriate level of subsidy is difficult and the numbers presented above are associated with substantial uncertainty and thus only provide the order of magnitude of the cost difference.

One way to avoid the problem of deciding the level of subsidy would be for EU to buy the ships at the market price, i.e. the price that breakers in Asia would be willing to pay. EU could then tender the dismantling of the vessels to the most competitive environmentally green facility in EU.

The cost difference would be the level of the subsidy and would have to be covered by EU. At least from a theoretical point of view this would not only be an efficient way to ensure green dismantling, it would also avoid the problem of explicitly deriving the level of subsidy needed.

As an alternative to direct subsidies, grants could be given to ship dismantling facilities to provide capacity, i.e. grants to cover investments.

Direct subsidies could potentially improve the present situation quickly by ensuring at least minimum "green" capacity for the dismantling of vessels. While establishment of capacity is of course a necessary condition, it is not a sufficient condition to ensure that ships are dismantled at the green facility.

Indeed, the most important cost driver is that of labour cost (an operating cost). Therefore, even when investments are covered for green facilities in EU they will not be competitive to the facilities with low labour cost in Asia. In a recent report by the French Secrétariat Général de la Mer (MIDN 2007) the costs of upgrading existing dismantling facilities in third world countries to a green recycling standard have been estimated to 50 to 150 USD per LDT (Annexe IX, p. 76).

Hence, upgrading and using existing Third World country facilities could be economically more attractive than using green facilities in Europe (which would still have higher costs in the range of 150 to 300 USD per LDT).

***Existing national strategies***

Subsidies to any industry should generally be avoided unless special circumstances apply. However, unsafe and environmentally unfriendly conditions of ship dismantling in Asia could justify a certain degree of subsidy for an interim period until world wide regulation ensuring acceptable dismantling is in place.

The Commission has recently approved an Italian state aid programme for the breaking-up of single hull oil tankers. The scheme promotes scrapping of small single hull oil tankers over fifteen years old operating on the Italian cabotage (local) market. Shipping companies will be granted aid to break up their single hull oil tankers over fifteen years old intended for the transport of crude oil or petroleum or chemical products, and no conditions are attached to the subsidy with respect to the conditions of the recycling facility.

The scheme will be in force for three years with an annual budget of €12 million. The ships in question will be demolished instead of being transferred to less strict non-Community registers.

In this way the measure anticipates the Community timetable and helps to improve shipping safety while protecting the marine environment. The Commission found that this measure boosts maritime transport safety without running counter either to the common interest or to European rules.

## UPGRADING FACILITIES OUTSIDE THE EU

One of the claims of the shipping industry regarding the recycling industry is that it is not transparent to the ship owners which yards are actually complying with standards of safe and environmental sound ship recycling.

One reason for this is of course that there is no coherent set of certifiable standards developed specifically for ship recycling, but instead in each country a number of regulations addressing worker's health, environmental monitoring, waste, waste water etc.

As stated above it may be appropriate to develop an internationally accepted base set of implementable and certifiable standards for green recycling facilities. If the IMO Convention eventually does make reference to standards for land-based facilities, these may be used already in the period up to entry into force for technical assistance projects, and if not, the guidance to the yards on how to comply with the demands of ship owners concerned with green recycling will be even more needed.

It is encouraging that a limited number of yards in Turkey, India and China have chosen to initiate a process towards green recycling under the prevailing market conditions and legislative environment, but very few have actually managed to buy vessels at a price reflecting the operating and capital costs of a green yard.

For the developing countries, the implementation of safe and environmentally sound ship recycling would necessitate investment in adequate waste reception facilities at the recycling yards, improved safety and health conditions for workers and environmentally sound waste management systems to ensure compliance with the appropriate recycling guidelines.

These investments are not likely to be driven by the market demand. One solution could be to develop and promote globally a programme with a strategy for resource mobilization from donors such as the European Development Bank, EurAid, UNDP, the World Bank, Asian Development Bank, the Global Environment Facility, other (inter) national donor organisations and bilateral donors to give technical assistance to the facilities in third world countries.

It is therefore suggested that the following is taken into consideration:

- Inclusion of the issue in bilateral agreements with relevant countries to bring attention to the matter.
- Technical assistance to the development and implementation of national roadmaps to improved conditions in breaker yards.
- Assistance to the shipping industry, e.g. with development of certification/ auditing schemes for green recycling facilities and lists of approved facilities.

The establishment of green recycling capacity for the merchant fleet is a task that must be driven primarily from the shipping and the recycling industry in response to the market, regulation or company policies. Obviously, the institutions of the EU may assist in capacity building, demonstration projects on upgrading in yards and trial vessels recycling.

Therefore, the successful promotion of sound recycling practices in ship breaking nations may require mandatory international regulation and/or programmes designed to assist capacity development.

**In EU candidate countries**

Pre-accession programmes exist for assisting these countries in achieving compliance with the EU *acquis*. These funds are prioritised by the countries and the Commission and could for Turkey be a significant and fast way if establishing the regulatory compliance of the local ship recycling industry.

The ongoing improvements in Aliaga were initiated by the Turkish Shipbreaker's Association and several yards, but authorities and industry appear committed to engage in activities to upgrade the local ship breaking industry.

**In India and China**

These countries do not accept assistance easily and do have the capacity to manage large investments. Assistance may be supplied through technical bilateral programmes. Recently, the authorities in Gujarat Province have finalised a hazardous waste disposal facility that may service Alang and a programme for ISO certification.

Thus, further capacity building for full environmental compliance should be well received. It should be noted that upgrading solely the environmental conditions of the yards will not result in acceptable conditions if the safety and health issues have not been addressed. No casualty statistics are available for ship recycling in China, but in Alang during 2003-2004 the official account of work-related fatal accidents were 26, although other sources claim there were 103 deaths due to a variety of accidents at the scrap yards.

The Accident Frequency Rate (AFR) could serve to measure the actual safety standard at the facilities, as a lower number of accidents can be related to the fact that fewer ships are being scrapped.

**Developing countries Bangladesh and Pakistan**

There has been very little scrapping activity until late 2006 and little if any donor assistance to the Pakistani yards. In Bangladesh the ILO has carried out a capacity building project in the ship breaking yards of Chittagong. Conditions in the yards have not improved over recent years and with the presently steaming recycling market in Bangladesh which is now the largest ship scrapping nation, more workers than ever are employed and more steel import levies are collected for the government.

This means that probably more that half a million people in extended families depend on the industry for their living. The consequences of a sudden unmitigated enforcement of regulations on the scrapping activities would be catastrophic in the areas engaged in the beach breaking today. With the upcoming peak in tanker phase-out volume Bangladesh will presumably be a prime destination for obsolete vessels.

Both Pakistan and Bangladesh are eligible for assistance from a number of donor sources, both bilateral and multilateral. Capacity building in terms of safe and environmentally sound management of ship recycling in yards and authorities may be needed and not least establishment of waste collection and hazardous waste facilities would help to advance the recycling conditions in these countries.

**Who should pay?**

As green recycling means extra costs this raises the question of who shall bear the costs for green recycling? Should it be tax payer financed or should the costs be paid by the ship builders or the ships owners? In the longer run it will properly not be acceptable for investment in clean dismantling of ships to be subsidised out of public funds.

Hence, the objective should properly be to establish a sustainable funding system whereby shipowners contribute at least partly to the safe and environmentally sound dismantling of ships worldwide. For government vessels, where no additional regulations are needed for green dismantling to take place at facilities in Europe/Turkey, the cost (in terms of costs difference to the most economic attractive option of beaching in Asia) is covered directly by the ship owner.

This follows a broad interpretation of the polluter pay principle as the industry itself finances the additional costs. If the economic instrument of subsidy is used for the merchant fleet, costs are first put on the tax payers via the EU and/or its member states. However, a subsidy scheme could be supplemented by a financing mechanism, which would generate funding from the industry.

Financing could potentially be provided using a number of different options:

- Levies on the shipping industry.
- Tax on calling at a European port.
- Recycling fund with contributions to be paid at the new built phase.

Levies and taxes on the shipping industry to provide finance and cover the costs of a subsidy scheme in order to ensure green recycling could come in many different forms. Levies could be put on the new building phase but it could also be contributions during the lifetime of the vessel. Thus, for example recycling charges could be included in the insurance premium or they could be levied by the flag state. Regardless of the form of taxes and levies, it is vital that a scheme is designed in such a way that it is not possible to escape by re-flagging of vessels or otherwise evading or bypassing the regulation.

Therefore, due to the strongly global nature of the shipping industry, a financing scheme would most efficiently be implemented on international level. If the EU adopts legislation on its own without support of the IMO, the range of possible efficient tax schemes is reduced significantly. For example levying the ship building industry would not be effective as construction could take place outside EU.

In other words, introduction of poorly designed financing mechanisms would lead to circumvention of the rules and thus have no effect in relation to the objective of providing finance.

## AIR MASSES AND FRONTS

### AIR MASS

In meteorology, an air mass is a volume of air defined by its temperature and water vapor content. Air masses cover many hundreds or thousands of square miles, and adopt the characteristics of the surface below them. They are classified according to latitude and their continental or maritime source regions. Colder air masses are termed polar or arctic, while warmer air masses are deemed tropical.

Continental and superior air masses are dry while maritime and monsoon air masses are moist. Weather fronts separate air masses with different density characteristics. Once an air mass moves away from its source region, underlying vegetation and water bodies can quickly modify its character. Classification schemes tackle an air mass' characteristics, and well as modification...

#### Classification and notation

The Bergeron classification is the most widely accepted form of air mass classification, though others have produced more refined versions of this scheme over different regions of the globe. Air mass classification involves three letters. The first letter describes its moisture properties, with c used for continental air masses and m for maritime air masses. The second letter describes the thermal characteristic of its source region: T for Tropical, P for Polar, A for arctic or Antarctic, M for monsoon, E for Equatorial, and S for superior air. The third letter is used to designate the stability of the atmosphere. If the air mass is colder than the ground below it, it is labeled k. If the air mass is warmer than the ground below it, it is labeled w.

Air masses of oceanic origin are denoted with a lower-case "m" ("maritime"), while air masses of continental origin are denoted with a lower-case "c" ("continental"). Air masses are also denoted as either Arctic, polar, tropical, or equatorial.

These two sets of attributes are used in combinations depending on the air mass being described. For instance, an air mass originating over the desert southwest of the United States in summer may be designated "cT". An air mass originating over northern Siberia in winter may be indicated as "cA". An upper case "S" was occasionally used to denote something called a "superior" air mass. This was regarded as an adiabatically drying and warming air mass descending from aloft. In South Asia, an upper case "M" has been occasionally used to denote an air mass within the summer monsoon regime in that region.

The stability of an air mass may be shown using a third letter, either "k" or "w". An example of this might be a polar air mass blowing over the Gulf Stream, denoted as "cPk". Occasionally, one may also encounter the use of an apostrophe or "degree tick" denoting that a given air mass having the same notation as another it is replacing is colder than the replaced air mass. For example, a series of fronts over the Pacific might show an air mass denoted mPk followed by another denoted mPk'.

Another convention utilizing these symbols is the indication of modification or transformation of one type to another. For instance, an Arctic air mass blowing out over the Gulf of Alaska may be shown as "cA-mPk". Yet another convention indicates the layering of air masses in certain situations. For instance, the overrunning of a polar air mass by an air mass from the Gulf of Mexico over the Central United States might be shown with the notation "mT/cP".

### Characteristics

Arctic, Antarctic, and polar air masses are cold. The qualities of arctic air are developed over ice and snow-covered ground. Arctic air is deeply cold, colder than polar air masses. Arctic air can be shallow in the summer, and rapidly modify as it moves equatorward. Polar air masses develop over higher latitudes over the land or ocean, are very stable, and generally shallower than arctic air. Polar air over the ocean loses its stability as it gains moisture over warmer ocean waters.

Tropical and equatorial air masses are hot as they develop over lower latitudes. Those that develop over land are drier and hotter than those that develop over oceans, and travel northward on the western periphery of the subtropical ridge. Maritime tropical air masses are sometimes referred to as trade air masses.

Monsoon air masses are moist and unstable. Superior air masses are dry, and rarely reach the ground. It normally resides over maritime tropical air masses, forming a warmer and drier layer over the more moderate moist air mass below, forming what is known as a trade wind inversion over the maritime tropical air mass.

Continental Polar air masses are air masses that are cold and dry due to their continental source region. Continental polar air masses that affect North America form over interior Canada. A Continental Tropical Air Mass is a type of tropical air produced over subtropical arid regions; it is hot and very dry.

**Movement and fronts**

A weather front is a boundary separating two masses of air of different densities, and is the principal cause of meteorological phenomena. In surface weather analyses, fronts are depicted using various colored lines and symbols, depending on the type of front.

The air masses separated by a front usually differ in temperature and humidity. Cold fronts may feature narrow bands of thunderstorms and severe weather, and may on occasion be preceded by squall lines or dry lines. Warm fronts are usually preceded by stratiform precipitation and fog. The weather usually clears quickly after a front's passage. Some fronts produce no precipitation and little cloudiness, although there is invariably a wind shift.

Cold fronts and occluded fronts generally move from west to east, while warm fronts move poleward. Because of the greater density of air in their wake, cold fronts and cold occlusions move faster than warm fronts and warm occlusions. Mountains and warm bodies of water can slow the movement of fronts. When a front becomes stationary, and the density contrast across the frontal boundary vanishes, the front can degenerate into a line which separates regions of differing wind velocity, known as a shearline. This is most common over the open ocean.

**Modification**

Air masses can be modified in a variety of ways. Surface flux from underlying vegetation, such as forest, acts to moisten the overlying air mass. Heat from underlying warmer waters can significantly modify an air mass over distances as short as 35 kilometres to 40 kilometres.

For example, southwest of extratropical cyclones, curved cyclonic flow bringing cold air across the relatively warm water bodies can lead to narrow lake-effect snow bands. Those bands bring strong localized precipitation since large water bodies such as lakes efficiently store heat that results in significant temperature differences between the water surface and the air above. Because of this temperature difference, warmth and moisture are transported upward, condensing into vertically oriented clouds which produce snow showers. The temperature decrease with height and cloud depth are directly affected by both the water temperature and the large-scale environment.

The stronger the temperature decrease with height, the deeper the clouds get, and the greater the precipitation rate becomes.

## TEMPERATURE AND TROPICAL CYCLONES

A tropical cyclone is composed of a system of thunderstorms that shows a cyclonic rotation around a central core or eye. A *tropical cyclone* is a generic term for a storm with an organized system of thunderstorms that are not based on a frontal system.

Each individual tropical cyclone differs, but several characteristics are common to most all tropical cyclones including a central low-pressure zone and high wind speeds of at least 34 knots.

At this point, the storms are given a pre-determined storm name. Most storms are accompanied by a lot of rain and storm surges near the shore. Often, once the storms make landfall, the tropical cyclone can cause tornadoes.

A tropical cyclone needs warm ocean temperatures in order to form. Temperatures in the ocean need to be at least 82 degrees Fahrenheit in order to form. Heat is drawn up from the oceans creating what is popularly called a 'heat engine'.

Tall convective towers of clouds are formed within the storm as warm ocean water evaporates. As the air rises higher it cools and condenses releasing latent heat which causes even more clouds to form and feed the storm.

### ROTATION AND FORWARD SPEED

The rotation of tropical cyclones in the Northern Hemisphere is counter-clockwise due to the Coriolis Effect. The opposite is true in the Southern Hemisphere.

The forward speed of a tropical cyclone can be a factor in determining the amount of damage the storm will cause. If a storm remains over one area for a long period of time, torrential rains, high winds, and flooding can severely impact an area.

The average forward speed of a tropical cyclone is dependent on the latitude where the storm is currently. Generally, at less than 30 degrees of latitude, the storms will move at about 20 mph on average. The closer the storm is located the the equator, the slower the movement. Some storms will even stall out over an area for an extended period of time. After about 35 degrees North latitude, the storms start to pick up speed.

A good example of the fast formation of tropical cyclones comes when several storms stack up in the ocean back-to-back. Such an example occurred in 2009 with the formation of Ana, Bill, and Claudette as seen in this satellite image. The storms were very close to one another. Storms can also become entangle with one another in a process known as the Fujiwhara Effect where tropical cyclones can interact with each other.

## NAMING TROPICAL CYCLONES

In the Atlantic Ocean and Eastern Pacific, tropical cyclones are known as hurricanes. In the Western Pacific Ocean, tropical cyclones are known as typhoons. In the Indian Ocean, a tropical cyclone is called a cyclone. These names are described in the article – Is it a typhoon, a cyclone, or a hurricane?

Specific storm names in each of the ocean basins vary based on conventional naming practices. For instance, in the Atlantic Ocean, storms are given names based on an alphabetical pre-determined list of Atlantic hurricane names. Severe hurricanes names are often retired.

## TYPES AND DISTRIBUTION OF PRECIPITATION

The global distribution of precipitation is influenced by the general circulation of the atmosphere, proximity to large bodies of water, and topography. Precipitation is most abundant where air rises, and least abundant where it sinks. It also tends to be greater near oceans and lakes, and in higher elevations.

## TERTIARY CIRCULATION

Tertiary circulations are localized circulations directly attributable to one of the following causes or a combination of them: local cooling, local heating, adjacent heating or cooling, and induction. Many regions have local weather phenomena caused by temperature differences between land and water surfaces or by local topographical features. These weather phenomena show up as circulations.

These tertiary circulations can result in dramatic local weather conditions and wind flows. The most common tertiary circulations are discussed in this lesson. However, there are numerous other circulations and related phenomena in existence around the world.

## MONSOON WINDS

The term *monsoon* is of Arabic origin and means season. The monsoon wind is a seasonal wind that blows from continental interiors to the ocean in the winter; they blow in the opposite direction during the summer. The monsoon wind is most pronounced over India, although there are other regions with noticeable monsoon winds.

Monsoon winds are a result of unequal heating and cooling of land and water surfaces. During winter a massive area of cold high pressure develops over the extensive Asiatic continent.

This high pressure is due primarily to cold arctic air and long-term radiation cooling. To the south, the warm equatorial waters exist and, in contrast, the area has relatively lower surface pressures. The combination of high pressure over Asia and low pressure over the Equatorial Belt sets up a pressure gradient directed from north to south. Because of the flow around the massive Siberian high, northeast winds begin to dominate the regions from India to the Philippines.

During the winter months, clear skies predominate over most of the region. This is caused by the mass motion of air from a high-pressure area over land to an area of lower pressure over the ocean. As the air leaves the high-pressure area over land, it is cold and dry.

As it travels over land toward the ocean, there is no source of moisture to induce precipitation. The air is also traveling from a higher altitude to a lower altitude; consequently, this downslope motion causes the air to be warmed at the adiabatic lapse rate. This warming process has a still further clearing effect on the skies. During the summer the airflow over the region is completely reversed. The large interior of Asia is heated to the point where the continent is much warmer than the ocean areas to the south. This induces relatively low pressure over Asia and higher pressure over the equatorial region.

The weather associated with the summer monsoon winds is thunderstorms, almost constant heavy rain, rain showers, and gusty surface winds. This condition is caused by mass motion of air from the relatively high-pressure area over the ocean to a low-pressure area over land. When the air leaves the ocean, it is warm and moist. As the air travels over land toward the low-pressure area, it is also traveling from a lower altitude to a higher altitude. The air is lifted by a mechanical force and cooled to its condensation point by this upslope motion.

## LAND AND SEA BREEZES

There is a diurnal contrast in the heating of local water and land areas similar to the seasonal variation of the monsoon. During the day, the land is warmer than the water area; at night the land area is cooler than the water area. A slight variation in pressure is caused by this temperature contrast. At night the wind blows from land to sea and is called a land breeze. During the day, the wind blows from water areas to land areas and is called a sea breeze.

The sea breeze usually begins during midmorning when the land areas become warmer than adjacent ocean waters. This temperature difference creates an area of slightly lower surface pressures over land compared to the now cooler waters. The result is a wind flow from water to land. The sea breeze starts with a shallow flow along the surface; however, as maximum heating occurs, the flow increases with height.

The height varies from an average of 3,000 feet in moderately warm climates to 4,500 in tropical regions. The effects of the sea breeze can be felt as far as 30 miles both onshore and offshore. In extreme cases, the sea breeze is felt 100 miles inland depending upon terrain. By mid afternoon the sea breeze will reach its maximum speed and may be strong enough to be influenced by the Coriolis force, which causes it to flow at an angle to the shore. The sea breeze is most pronounced in late spring, summer, and early fall when maximum temperature differences occur between land and water surfaces. A decrease in temperature and an increase in humidity and wind speed mark the start of a sea breeze.

The sea breeze continues until the land area cools sufficiently to dissipate the weak low pressure. After sunset, the land cools at a faster rate than the adjacent waters and eventually produces a complete reversal of the winds. As the land continues to cool through the evening hours, a weak area of high pressure forms over the land. The water area, with its warmer temperatures, has slightly lower pressure and again a flow is established; however, the flow is now from land to Water.

The land breezes, when compared to the sea breezes, are less extensive and not as strong. This is because there is less temperature contrast at night between land and water surfaces as compared to the temperature contrast during daytime heating.

Land breezes are at maximum development late at night, in late fall and early winter. In the tropical land regions, the land and sea breezes are repeated day after day with great regularity. In high latitudes the land and sea breezes are often masked by winds of synoptic features.

## WINDS DUE TO LOCAL COOLING AND HEATING

In the next sections we discuss tertiary circulations due to local cooling and heating effects. Under normal circumstances, these winds attain only light to moderate wind speeds; however, winds often occur in and near mountain areas that have undergone dramatic changes in normal character. At times, mountain areas tend to funnel winds through valleys and mountain passes. This funneling effect produces extremely dangerous wind speeds.

## FUNNEL EFFECT

Winds blowing against mountain barriers tend to flatten out and go around or over them. If a pass or a valley breaks the barrier, the air is forced through the break at considerable speed. When wind is forced through narrow valleys it is known as the funnel effect and is explained by Bernoulli's theorem. According to Bernoulli's theorem, pressures are least where velocities are greatest; likewise, pressures are greatest where velocities are least. This observation is true for both liquids and gases.

Bernoulli's theorem is frequently used to forecast tertiary winds in the mountainous western United States. The famous Santa Ana winds of southern California are a prime example. Winds associated with high pressure situated over Utah are funneled through the valley leading into the town of Santa Ana near the California coast.

Low pressure develops at the mouth of the valley and the end result is hot, dry, gusty and extremely dangerous winds. When the Santa Ana is strong enough, the effects are felt in virtually every valley located along the coast of southern California.

Visibility is often restricted due to blowing sand. It is common to see campers, trailers, and trucks turned over by the force of these winds. When funneled winds reach this magnitude, they are called jet-effect winds, canyon winds, or mountain-gap winds.

### *Winds Due to Local Cooling and Heating*

There are two types of tertiary circulations produced by local cooling—glacier winds and drainage winds.

#### *GLACIER WINDS*

Glacier winds, or fall winds occur in many varieties in all parts of the world where there are glaciers or elevated land masses that become covered by snow and ice during winter. During winter, the area of snow cover becomes most extensive.

Weak pressure results in a maximum of radiation cooling. Consequently the air coming in contact with the cold snow cools. The cooling effect makes the overlying air more dense, therefore, heavier than the surrounding air. When set in motion, the cold dense air flows down the sides of the glacier or plateau. If it is funneled through a pass or valley, it may become very strong. This type of wind may form during the day or night due to radiation cooling. The glacier wind is most common during the winter when more snow and ice are present.

When a changing pressure gradient moves a large cold air mass over the edge of a plateau, this action sets in motion the strongest, most persistent, and most extensive of the glacier or fall winds. When this happens, the fall velocity is added to the pressure gradient force causing the cold air to rush down to sea level along a front that may extend for hundreds of miles.

This condition occurs in winter on a large scale along the edge of the Greenland icecap. In some places along the icecap, the wind attains a velocity in excess of 90 knots for days at a time and reaches more than 150 nautical miles out to sea. Glacier winds are cold katabatic winds. Since all katabatic winds are heated adiabatically in their descent, they are predominantly dry.Occasionally, the glacier winds pick up moisture from falling precipitation when they underride warm air.

Even with the adiabatic heating they undergo, all glacier or fall winds are essentially cold winds because of the extreme coldness of the air in their source region. Contrary to all other descending winds that are warm and dry, the glacier wind is cold and dry. It is colder, level for level, than the air mass it is displacing. In the Northern Hemisphere, the glacier winds descend frequently from the snow-covered plateaus and glaciers of Alaska, Canada, Greenland, and Norway.

*DRAINAGE WINDS*

Drainage winds are caused by the cooling air along the slopes of a mountain. Consequently, the air becomes heavy and flows downhill, producing the MOUNTAIN BREEZE. Drainage winds are katabatic winds and like glacier winds; a weak or nonexistent pressure gradient is required to start the downward flow. As the air near the top of a mountain cools through radiation or contact with colder surfaces, it becomes heavier than the surrounding air and gradually flows downward.

Initially this flow is light and only a few feet thick. As cooling continues, the flow increases achieving speeds up to 15 knots at the base of the mountain and a depth of 200 feet or more. Winds in excess of 15 knots are rare and only occur when the mountain breeze is severely funneled. Drainage winds are cold and dry.

Adiabatic heating does not sufficiently heat the descending air because of the relative coldness of the initial air and because the distance traveled by the air is normally short. Drainage winds have a very localized circulation. As the cold air enters the valley below, it displaces the warm air. Temperatures continue to fall. If the flow achieves speeds of 8 knots or more, mixing results between the warm valley air and the cold descending air that results in a slight temperature increase. Campers often prefer to make summer camps at the base of mountains to take advantage of the cooling effect of the mountain breeze.

## MEASUREMENT OF TEMPERATURE

Temperature in the ocean has been measured many ways. Thermistors and mercury thermometers are commonly used on ships and buoys. These are calibrated in the laboratory before being used, and after use if possible, using mercury or platinum thermometers with accuracy traceable to national standards laboratories. Infrared radiometers on satellites measure the ocean's surface temperature.

### Mercury Thermometer

This is the most widely used, non-electronic thermometer. It was widely used in buckets dropped over the side of a ship to measure the temperature of surface waters, on Nansen bottles to measure subsea temperatures, and in the laboratory to calibrate other thermometers. Accuracy is about ± 0.001°C with careful calibration.

One very important mercury thermometer is the reversing thermometer carried on Nansen bottles, which are described in the next section. It is a thermometer that has a constriction in the mercury capillary that causes the thread of mercury to break at a precisely determined point when the thermometer is turned upside down. The thermometer is lowered deep into the ocean in the normal position; and it is allowed to come to equilibrium with the water. Mercury expands into the capillary, and the amount of mercury in the capillary is proportional to temperature. The thermometer is then flipped upside down, the thread of mercury breaks trapping the mercury in the capillary, and the thermometer is brought back. The mercury in the capillary of the reversed thermometer is read on deck along with the temperature of a normal thermometer, which gives the temperature at which the reversed thermometer is read. The two readings give the temperature of the water at the depth where the thermometer was reversed.

The reversing thermometer is carried inside a glass tube which protects the thermometer from the ocean's pressure because high pressure can squeeze additional mercury into the capillary. If the thermometer is unprotected, the apparent temperature read on deck is proportional to temperature and pressure at the depth where the thermometer was flipped A pair of protected and unprotected thermometers gives temperature and pressure of the water at the depth the thermometer was reversed.

Pairs of reversing thermometers carried on Nansen bottles were the primary source of subsea measurements of temperature as a function of pressure from around 1900 to 1970.

## THE OCEANIC MIXED LAYER AND THERMOCLINE

Wind blowing on the ocean stirs the upper layers leading to a thin mixed layer at the sea surface having constant temperature and salinity from the surface down to a depth where the values differ from those at the surface. The magnitude of the difference is arbitrary, but typically the temperature at the bottom of the layer must be no more than 0.02-0.1°C colder than at the surface.

Note that the both temperature and salinity must be constant in the mixed layer. The mixed layer is roughly 10-200 m thick over most of the tropical and mid-latitude belts. The depth and temperature of the mixed layer varies from day to day and from season to season in response to two processes:

1. Heat fluxes through the surface heat and cool the surface waters. Changes in temperature change the density contrast between the mixed layer and deeper waters. The greater the contrast, the more work is needed to mix the layer downward and visa versa.
2. Turbulence in the mixed layer mixes heat downward. The turbulence depends on the wind speed and on the intensity of breaking waves. Turbulence mixes water in the layer, and it mixes the water in the layer with water in the thermocline.

The mid-latitude mixed layer is thinnest in late summer when winds are weak, and sunlight warms the surface layer. At times, the heating is so strong, and the winds so weak, that the layer is only a few meters thick. In fall, the first storms of the season mix the heat down into the ocean thickening the mixed layer, but little heat is lost. In winter, heat is lost, and the mixed layer continues to thicken, becoming thickest in late winter. In spring, winds weaken, sunlight increases, and a new mixed layer forms.

Below the mixed layer, water temperature decreases rapidly with depth except at high latitudes. The range of depths where the rate of change, the gradient of temperature, is large is called the thermocline. Because density is closely related to temperature, the thermocline also tends to be the layer where density gradient is greatest, the pycnocline. The top of the thermocline varies slightly with the seasons. This is the seasonal thermocline. The permanent thermocline extends from below the seasonal thermocline to depths of 1500-2000m.

The mixed layer tends to be saltier than the thermocline between 10° and 40° latitude, where evaporation exceeds precipitation. At high latitudes the mixed layer is fresher because rain and melting ice reduce salinity. In some tropical regions, such as the warm pool in the western tropical Pacific, rain also produces a thin fresher mixed layer.

# Index

## K



## M



## N



## O



## P



## R



## S



## T



## V



## W



## Y